PREDICTIONS

*Society's Telltale Signature
Reveals the Past
and Forecasts the Future*

Theodore Modis, Ph.D.

Simon & Schuster
New York   London   Toronto   Sydney   Tokyo   Singapore

SIMON & SCHUSTER
Simon & Schuster Building
Rockefeller Center
1230 Avenue of the Americas
New York, New York 10020

SIMON & SCHUSTER and colophon are registered trademarks
of Simon & Schuster Inc.

Designed by Irving Perkins Associates
Manufactured in the United States of America

1   3   5   7   9   10   8   6   4   2

Library of Congress Cataloging-in-Publication Data

Modis, Theodore.
    Predictions : society's telltale signature reveals the past
and forecasts the future / Theodore Modis.
        p.   cm.
    Includes bibliographical references and index.
    1. Forecasting. 2. Creation (Literary, artistic, etc.)
3. Science and civilization. I. Title.
CB 158.M63   1992
303.49—dc20   92-18321
                                                        CIP

ISBN: 0-671-75917-5

To my mother

CONTENTS

PROLOGUE   11

Spicy Talk and a Stimulating Meal   13

1. SCIENCE AND FORETELLING   19

Invariants   23

    Safety in Cars   24

    Rationing Computer Innovation   26

    Humans Versus Machines   27

    Do All Animals Die at the Same Age?   30

The Bell-Shaped Curve of the Life Cycle   32

The S-Curve of Cumulative Growth   33

The Law of Natural Growth   35

    Controls on Rabbit Populations   36

Gauss for Statistics Was Like Bach for Music—an Ingenious Forger   38

2. NEEDLES IN A HAYSTACK   41

How Much Order Is There in ''Random'' Discoveries?   41

Learning from Success   44

    Could You Take Aim at an Apple on Your Son's Head?   45

    A Real Case   46

    The Industry Learning Curve   48

Going West   49

Mesopotamia, the Moon, and the Matterhorn   51

3. INANIMATE PRODUCTION LIKE ANIMATE REPRODUCTION   55
   Do Not Try to Forecast What You Already Know   61
   Man-made Dinosaurs   63
   Mama Earth   67
   Did Christianity Begin Before Christ?   70

4. THE RISE AND FALL OF CREATIVITY   73
   Mozart Dies of Old Age   75
   Did Einstein Publish Too Much?   78
   Can We Predict Our Own Death?   79
   Searching for Failures of the S-Curve Approach   81
        An Unnatural Suicide Attempt   83
   A "Fertility" Template   85
   Contemporary Geniuses   86
   Innovation in Computer Making   87

5. GOOD GUYS AND BAD GUYS COMPETE THE SAME WAY   91
   Criminal Careers   95
   Killer Diseases   97
   When Will There Be a Miracle Drug for AIDS?   100

6. A HARD FACT OF LIFE   106
   Straight Lines in Nature   111
   The Cultural Resistance to Innovation   116
   Unnatural Substitutions   118
   The Swedes Are Coming   120
   Fishing and Prying   122

7. COMPETITION IS THE CREATOR AND THE REGULATOR   124
   When It Rains, It Pours   127
   Successive Substitutions   133

Only One Front-Runner at a Time   135

Transport Infrastructures   136

The Primary Energy Picture   138

The Disease "Market"   143

Nobel Prize Awards   144

8. A COSMIC HEARTBEAT   147

The Wall Street Crash of 1987   155

Matters of Life and Death   157

Epitaphs   162

9. REACHING THE CEILING EVERYWHERE   166

Clocking the Transport of the Future   176

Futuronics   181

    On the Possibility and Desirability of Peace   182

    Global Village in 2025   185

10. IF I CAN, I WANT   188

What Makes a Tourist?   191

The Nutritional Value of New Impressions   192

From Tourism to S-Curves   193

From S-Curves to Chaos   195

    Before and After Chaos   200

11. FORECASTING DESTINY   205

Decision Non-makers   208

A System's Predictability   215

    Wasteful Computers   216

Little Choice   218

EPILOGUE   223

APPENDIX A. MATHEMATICAL FORMULATION OF S-CURVES
      AND THE PROCEDURE FOR FITTING THEM ON DATA   229
The Predator-Prey Equations   229
The Malthusian Case: One Species Only   230
One-to-One Substitutions: Two Competitors in a Niche   231
Multiple Competition: Many Competitors at the Same Time   232
Fitting an S-Curve on a Set of Data Points   232

APPENDIX B. EXPECTED UNCERTAINTIES ON S-CURVE
      FORECASTS   234

APPENDIX C. ADDITIONAL FIGURES   237

NOTES AND SOURCES   275

ACKNOWLEDGMENTS   285

INDEX   287

PROLOGUE

The fisherman starting his day off the Adriatic coast was wondering whether it was going to be a day of big fish or small ones. He had seen this phenomenon often. He would start by catching a big fish in the morning and then for the rest of the day it would be one big fish after another. Other times it would be small catches all day long. He was reminded in passing of a biblical reference to periods of fat years and thin years, but he got down to work without further philosophizing. He had no time to waste; in the days following the Great War the sea was one place where food could still be found relatively easily.

· · ·

Meanwhile, at the University of Siena, the biologist Umberto D'Ancona was making statistical studies of Adriatic fisheries. He found temporary increases in the mean relative frequency of the more voracious kinds of fish, as compared with the fish on which they preyed. Vito Volterra, a mathematician at the University of Rome, was preoccupied in his own way with the same phenomenon. He knew of D'Ancona's observations and believed he understood the reason for them. Since big fish eat small fish—and consequently depend on them for survival—some interchange in population dominance should be expected. The population of big fish would grow until small fish became scarce. At that point the big fish would starve and their diminishing numbers would give the small-fish survivors a chance to renew their numbers. Could this phenomenon be described mathematically?

Volterra succeeded in building a mathematical formulation that described well the fisherman's observations. A model for the growth of populations, it states that the rate of growth is limited by competition and that the overall size of the population (for example, the number of rabbits in a fenced-off grass field) slowly approaches a ceiling, the height of which reflects the capacity of the ecological niche. The model would serve as a foundation for modern biological studies of the competitive struggle for life. Alfred J. Lotka also studied such problems to some extent. Today, there are applications bearing both men's names.

Half a century later, Cesare Marchetti, a physicist at the International Institute of Advanced Systems Analysis (IIASA) near Vienna, Austria, was given the task by the energy-project leader to forecast energy demands. Another kind of war had been shaking the West recently: the fierce competition for oil. The need for increased understanding of the future energy picture was becoming imperative. Marchetti approached the problem as a physicist; he sought answers through the use of the scientific method: observation, prediction, verification. In this approach, predictions must be related to observations through a theory resting on hypotheses. When the predictions are verified, the hypotheses become *laws*. The simpler a law, the more fundamental it is and the wider its range of applications.

Marchetti had long been concerned with the "science" of predictions. In his work he first started searching for what physicists call *invariants*. These are constants universally valid and manifested through indicators that do not change over time. He believed that such indicators represent some kind of equilibrium, even if one is not dealing with physics but with human activities instead. He then suspected that the fundamental laws which govern growth and competition among species may also describe human activities. Competition can be as fierce in the marketplace as in the jungle, and the law of the survival of the fittest becomes indisputable. Marchetti noted that growth curves for animal populations follow patterns similar to those for product sales. Could it be that the mathematics developed by Volterra for the growth of a rabbit population describe equally well the growing numbers of cars and computers? Marchetti went on to make a dazzling array of predictions, including forecasts of future energy demands, using Volterra's equations. But how far can the analogy between natural laws and human activities be pushed, and how trustworthy are the quantitative forecasts based on such formulations?

Some years ago my professional lifeline crossed those of Volterra and Marchetti. I was moving from academia to industry, leaving fifteen years of research in elementary particle physics to work as a management science consultant for a prominent computer manufacturer. My boss, an ex-physicist himself, tried to smooth the transition by showing me some of Marchetti's papers that described applications of laws from the natural sciences to a variety of human activities. "See how we are also intellectually alert in industry" was the message. However, three weeks later, and in spite of my enthusiasm, the stern new message was: "Now leave all this aside and let's get down to work." It was too late, because the subject had intrigued me.

A few months later, I was asked to forecast the life cycles of computer products and the rate at which recent models substitute for older ones. Like Marchetti, I found myself in the position of being a physicist in the business of predicting the future. My reaction was immediate. As soon as I was charged officially with the forecasting project, I took the first plane for Vienna to see Marchetti at IIASA.

SPICY TALK AND A STIMULATING MEAL

Laxenburg, Austria, is a tiny old town that has stayed in the past even more than the nearby capital, Vienna. IIASA is housed in a chateau with a fairy-tale courtyard in which Cesare Marchetti strolls. His silver hair, spectacles, Tyrolean feathered hat, and Austrian cape make him the classic image of a venerated Austrian professor.

Not at all! An Italian, he received his doctorate in physics from the University of Pisa, and his activities evolved in a Leonardo da Vinci tradition to cover a wide area both in subject matter and geography. In the 1950s he worked on the technology of heavy water separation for the Italians. Afterward he represented EURATOM (the European Atomic Energy Commission) in Canada for two years. In the 1960s he researched nuclear reactor design and nuclear waste disposal as the head of the materials division of the European Community Research Center. In 1974 he retired and joined IIASA to study global energy systems.

Two years earlier, in 1972, a book appeared, *The Limits to Growth,* published by the Club of Rome, an informal international association with about seventy members of twenty-five nationalities.[1] Scientists, educators, economists, humanists, industrialists, they were all united by their conviction that the major problems of mankind today are too complex to be tackled by traditional institutions and policies. Their book drew alarming conclusions concerning the earth's rampant overpopulation and the depletion of raw materials and primary energy sources. Its message delivered a shock and contributed to the "think small" cultural wave of the 1970s.

Marchetti had persistently refused to join the Club of Rome on grounds of "self-respect," implying that it had failed to stay close to the fundamentals of science. His response to *The Limits to Growth* was an article titled "On 10^{12}: A Check on Earth Carrying Capacity for Man," written for his "friends of the Club of Rome."[2] In this article he provided calculations demonstrating that it is possible to sustain one trillion people on the earth without exhausting any basic resource, including the environment! It was one more brush stroke in his self-portrait as a maverick.

In 1985 I arrived at Laxenburg, seeking Marchetti at IIASA. He received me in his office buried behind piles of paper. Later I discovered that most of the documents contained numbers, the data sources serving him very much like sketchbooks serve artists. He welcomed me warmly, we uncovered a chair for me, and I went straight to the point. I showed him my first attempts at determining the life cycles of computers. I had dozens of questions. He answered laconically, simply indicating the direction in which I should go to search for my own conclusions. "Look at all computers together," he said. "Small and big ones are all competing together. They are filling the same niche. You must study how they substitute for one another. I've seen cars, trains, and other human creations go through this process."

What he said produced an echo in me. People can spend their money only once, on one computer or on another one. Bringing a new computer model to market depresses the sales of an older model. Also, one can type on only one keyboard at a time. A new keyboard with ergonomic design will eventually replace all flat keyboards in use. This process of substitution is a phenomenon similar to the growth of an animal population and follows the fundamental law of nature described by Volterra.

My discussion with Marchetti lasted for hours. He defended his arguments in a dogmatic way that often sounded arrogant. Some of the things he said were so provocative that it was difficult for me not to argue. I mostly kept quiet, however, trying to absorb as much as possible. During lunch he kept tossing out universal constants—what he called *invariants*—as if to add spice to our meal. Did I know that human beings around the world are happiest when they are on the move for an average of about seventy minutes per day? Prolonged deviation from this norm is met with discomfort, unpleasantness, and rejection. To obscure the fact that one is moving for longer periods, trains feature games, reading lounges, bar parlors, and other pastime activities. Airlines show movies during long flights. On the other hand, lack of movement is equally objectionable. Confinement makes prisoners pace their cell in order to meet their daily quota of travel time.

Did I know, Marchetti asked, that during these seventy minutes of travel time, people like to spend no more and no less than 15 percent of their income on the means of travel? To translate this into biological terms, one must think of income as the social equivalent of energy. And did I know that these two conditions are satisfied in such a way as to maximize the distance traveled? Poor people walk, those better off drive, while the rich fly. From African Zulus to sophisticated New Yorkers, they are all trying to get as far as possible *within the seventy minutes and the 15 percent budget allocation.* Affluence and success result in a bigger radius of action. Jets did not shorten travel time, they simply increased the distance traveled.

Maximizing range, Marchetti said, is one of the fundamental things that all organisms have been striving for from the most primitive forms to mankind. Expanding in space as far as possible is what reproducing unicellular amoebas are after, as well as what the conquest of the West and space exploration were all about. In his opinion, every other rationalization tends to be poetry.

On my return flight home I experienced a euphoric impatience. I took out my pocket notebook to organize my thoughts and jot down impressions, conclusions, and action plans. I decided to go straight to my office upon my arrival late Friday night. I had to try Volterra's equations to track the substitution of computers. Marchetti had said that looking at the whole market niche would reveal the detailed competition between the various models. Could this approach describe the substitution of computers as well as it described the many substitutions Marchetti spoke of? Could it be that replacing large computers by smaller ones is a "natural" process that can be quantified and projected far into the future? Would that be a means of forecasting the future of my company? Could the life cycles of organizations be predicted like those of organisms, and if so, with what accuracy? Would it even be possible to derive an equation for myself and estimate the time of my death?

Beyond my excitement, there was some suspicion. I did not know how much I could trust Marchetti. I had to check things out for myself. If there was a catch, it should become obvious sooner or later. The one thing I did trust, even if I was no longer in physics, was the scientific method.

In a few months most of my friends and acquaintances knew of my preoccupation with using Volterra's formulations and Marchetti's approach as a means of probing the future. Contrary to my experience as a particle physicist, interest in my work was now genuine. It was no longer the befuddlement of those who believe themselves intellectually inferior and admire the interest that must be there in something too difficult to understand. Now it was more like "This is really interesting! Can *I* make it work for *me?*"

Everyone who knew of my work was intrigued with the possibility that forecasts could become more reliable through the use of the natural sciences. Requests for more information, explanations, and specific applications kept pouring in from all directions. A friend who had just begun selling small sailboats wanted to know next summer's demand on Lake Geneva. Another, in the restaurant business, worried about his diminishing clientele and was concerned that his cuisine was too specialized and expensive compared to his competition. A depressed young woman was anxious to know when her next love affair would be, and a doctor who had had nine kidney stones in fifteen years wanted to know if and when there would be an end to this painful procession.

Besides those eager to believe in the discovery of a miraculous future-predicting apparatus, there were also the skeptics. They included those who mistrusted successful forecasts as carefully chosen successes among many failures; those who argued that if there was a method for foretelling the future, one would not talk about it but get rich on it instead; and those who believed that their future could not be predicted by third parties because it lay squarely in their own hands.

For my part, I soon reached a stage of agonizing indecision. On one hand, Marchetti's approach appealed to me. The scientific method and the use of biological studies on natural growth processes and competition inspired confidence. On the other hand, I needed to make my own checks and evaluate the size of the expected uncertainties. I also had to watch out for the trap of losing focus by indulging in the mathematics. But if in the end I confirmed an increased capability to forecast, how would I accommodate the predeterminism that it implied? The belief that one is able to shape one's own future clashed with the fascination of discovering that society has its own ways, which can be quantified and projected reliably into the future. The old question of free will would not be easily resolved.

I searched for more and more cases in which social growth processes fit the description of natural and biological ones. I also searched for discrepancies and carried out extensive simulations to understand the failures. I had embarked upon what I came to call my S-curve adventure, and it took me in surprising directions. Along the way I became convinced that there is a wisdom accessible to everyone in some of the scientific formulations that describe natural growth processes, and when they are applied to social phenomena, they make it possible to interpret and understand the past as well as forecast the future. Furthermore, I learned to visualize most social processes through their life cycles without resorting to mathematics. Such a visualization, I found, offered new perspectives on both the past and the future.

From my experience with S-curves over the past six years I have come to two realizations. The first one deals with the fact that many phenomena go through a life cycle: birth, growth, maturity, decline, and death. Time scales vary, and some phenomena may look like revolutions while others may look like natural evolutions. The element in common is *the way* in which the growth takes place; for example, things come to an end slowly and continuously, not unlike the way they came into existence. The end of a life cycle, however, does not mean a return to the beginning. The phases of natural growth proceed along S-curves, cascading from one to the next, in a pattern that reinvokes much of the past but leads to a higher level.

My second realization concerns predictability. There is a promise implicit in a process of natural growth, which is guaranteed by nature: The growth cycle will not stop halfway through. Whenever I come across a fair fraction of a growth process, be it in nature, society, business, or my private life, I try to visualize the full life cycle. If I have the first half as given, I can predict the future; if I am faced with the second half, I can deduce the past. I have made peace with my arrogance and have grown to accept that a certain amount of predetermination is associated with *natural* processes, as if by definition from the word natural.

I am not the first one to be impressed with the amount of predictability accessible through natural growth formulations. Besides the scientists I mentioned earlier, many others have written professional works on this subject. The goal I set for myself in writing this book was to share my experience with a wider public by showing, in simple terms, the impact this way of thinking about the past and the future can have on our everyday lives.

1

Science and Foretelling

A barman asks Andy Capp which one he would choose—money, power, happiness, or the ability to foretell the future.

"Foretell the future," Andy answers. "That way I can make money. Money will bring me power, and then I'll be happy!"

• • •

The dream of being able to foretell the future is as old as human nature. From antiquity, that dream found expression through Mediterranean oracles, the Chinese I Ching, and the tarot, all of which purported to predict future events but which, in fact, were primarily useful as means for expressing wise subconscious judgments. Throughout the centuries, foretelling the future was associated with religion, mysticism, magic, and the supernatural. The advice of fortune-tellers was eagerly sought, and practitioners were revered if their predictions came true, reviled if they did not. The Chinese court astrologers Hsi and Ho were decapitated for not having predicted the solar eclipse of October 22, 2137 B.C. At other times, those who claimed to be able to see into the future were burned at the stake as witches.

Today, astrology and horoscopes serve the same purpose, capitalizing on the human need to know the future, just as religion exploits the fear of death. Unlike religion, however, the goal is neither salvation nor

19

internal peace. Most often the desire to know the future focuses on material or personal gains. People may not openly admit it, but the ability to foretell the future means acquiring power, becoming superhuman.

Scientists normally disdain the notion that they, too, are in the business of fortune-telling, which is curious because science itself revolves around methodologies for telling the future. Scientists merely use a different vocabulary. In contrast to fortune-tellers, they talk about calculations instead of predictions, laws instead of fate, and statistical fluctuations instead of accidents. Yet the aim of the scientific method is the same. From the observation of past events, scientists derive laws that, when verified, enable them to predict future outcomes.

This is not to diminish the value of science. On the contrary, considering its impressive success, I want to emulate it. The desire to involve science in foretelling the future is not new. Science carries authority and respect. The word scientific is used as a synonym for proven. Various disciplines seek to be classified among the sciences to raise their status. We have social science, behavioral science, computer science, environmental science, management science, decision science, marketing science, organization science, administration science. Before long we may see business science, sports science, or love science. It sometimes seems, in fact, that having science as a surname may be an indication of being poorly endowed by it.

Weather forecasting has called on the hard sciences repeatedly over the years. During the 1950s and the 1960s the advent of digital computers and space satellites made global weather forecasting a promising candidate as a new "science." The twenty years that followed saw vast and expensive bureaucracies built up around weather forecasting centers on both sides of the Atlantic. Equipped with the most powerful computers, these places attracted resources and talent in the name of long-range weather prediction. Yet complicated models involving the solution of multiple equations and handling large amounts of data produced forecasts that were useful for only a few days. Thousands of talented-man years and billions of dollars later, James Gleick concluded in his book *Chaos* that "long-range weather forecasting is doomed."[1]

Economic forecasting masquerading as a "science" also involved elaborate econometric models with complicated and sometimes arbitrary equations aimed at predicting future trends from variables such as prices, money supply, and interest rates. If the resulting forecasts turned out to be absurd, the equations were modified until the prediction fell within accepted limits. One justification for persevering in this direction is the voluminous number of publications—methods, models, forecasts—that appear to be the essential measures of academic productivity. Another is the need of government and industry planners for a more "scientific" way of estimating future trends. Yet again, a large-scale effort has been invested but has produced, as a rule, economic forecasts that failed dismally.

Meteorologists and economists, with their entourage of programmers, were not the only forecasters to claim scientific credentials. Statisticians and mathematicians have also been involved, and the forecasting gurus of today are found in business schools. They are the masters of the "Time Series" approach. A Time Series is a sequence of historical data points at regular time intervals: daily temperature readings, stock market prices, annual death rates due to traffic accidents, and so forth. Computers are employed to crunch these numbers, separating the Time Series into a trend, a cyclical component, a seasonal component, and a remaining "unexplained variation," from which can be derived a "forecast."

Time Series forecasters perform this feat of statistical and mathematical magic without asking what it is they are looking at. They pride themselves on analyzing the data without introducing human biases—for example, *wanting* the stock to go up—to such an extent that they develop "expert systems," programs capable of processing the data and making the prediction without human intervention. One simply types in the sequence of historical numbers and out comes the forecast, complete with a full report.

What Time Series forecasters achieve, in effect, is automation. Anybody anywhere can type the numbers into a computer, hit the ENTER key, and obtain the results. No evaluation, understanding, or even thinking is necessary. Claiming kinship to artificial intelligence, Time Series packages treat temperature variations and cumulative sales alike, allowing ups *and* downs, and may well predict a future decrease in a child's height. Spyros Makridakis, a Time Series old-timer, carried out a lengthy study to arrive at the conclusion that "simple statistical methods in Time Series analyses perform at least as well as complicated ones."[2]

In my experience the simplest and best Time Series approach is to graph the data and extrapolate their form by eye. At any rate, the endeavor—extrapolating a series of data points without understanding their shape—can shed light on only the very near future. It does little more than the naive "meteo" method: Tomorrow will be the same as today. Time Series models perform no better on long-range forecasts than those of economists and meteorologists. Using science this way has proved of little help.

The scientists who have systematically shied away from forecasting are physicists. Until recently the best of them, attracted by glittering particle physics, swarmed around nuclear physics laboratories, shunning meteorology and economics as pseudo-sciences. Some venerated theorists openly admitted physics' inability to make predictions about phenomena as complicated as turbulence, which is extremely sensitive to initial conditions. In Gleick's Butterfly Effect, a butterfly stirring the air today in Peking can produce storms next month in New York. In a science fiction story, a time traveler steps on an insect millions of years ago and upon coming back to the present finds an entirely different world. The awesome complexity of such systems turns away physicists who seek elegant solutions.

Classical physics is excellent at describing billiard ball movements. The difficulty comes from putting many balls together (many, for physicists, can be anything greater than three). Molecules in a volume of gas behave very much like billiard balls, but there are too many of them and they bounce too often. Thermodynamics, the branch of physics that studies gases, makes predictions by focusing on the macroscopic global variables only: temperature, pressure, and volume. The bottom-up approach, tracking individual molecules, taxed the ingenuity of the best minds in physics for at least a hundred years and has served only in understanding, corroborating, and justifying the relations established experimentally between the overall variables.

Squeezed by diminishing returns, physicists have started to scatter outside the musty dungeons of particle accelerators in recent years. They are showing up in unlikely places: neurophysiology, Wall Street, chaos studies, and forecasting, to name a few. They carry with them detectors, tools, techniques, and tricks, all built around the scientific method: observation, prediction, verification. They are striving for understanding the future in the physics tradition—like hound dogs, sniffing out the right direction as they move along.

Derek J. de Solla Price starts as an experimental physicist but soon develops a passion for applying his training to scientometric problems (quantitative studies on the scientific community). He is the first to make the connection between natural growth and chaos. He spends much of his career as Avalon professor of history of science at Yale and deserves the characterization of being the Francis Galton* of the historiography and the sociology of science.

---

* Francis Galton (1822–1911) applied scientific skills on social problems as early as the mid-19th century. He was the grandson of Erasmus Darwin and among his achievements

Independently, Elliott Montroll, a solid-state physicist at the University of Rochester, discusses mathematical models for social phenomena. He writes about how competition affects traffic and population growth. He even suggests that money is transferred from one individual to another in a way analogous to that in which energy is transferred from gas molecule to gas molecule by collisions. Through the transfer of goods and services everyone has some annual income. One might expect that after many transactions money will be randomly distributed, but in fact, some people end up with significantly larger incomes than others.[4] Still, Montroll takes advantage quantitatively of some laws of thermodynamics.

Similarly and at about the same time, Cesare Marchetti looks at an energy system as composed of many small parts (individuals) in random motion. The observable is an envelope, such as overall energy consumption for the world, for a nation, or for one particular kind of fuel. To emulate physics one may want to reconstruct the envelope from the microscopic variables, the individual energy consumers. But this would be as difficult as it is for billiard balls and gas molecules. The proof lies with the failure of weather and economic forecasting attempts due to the vast number of microscopic variables.

A far wiser approach is to concentrate on the macroscopic variables that describe the overall behavior of a phenomenon. There is evidence to show that "individuals" within a system "arrange" their behavior so that such macroscopic descriptions as invariants and natural-growth functions bring out the system's *holographic* character. Thus, from a small part of history one is able to reconstruct successfully a long series of data points forward and backward in time. In other words, there are stable and deep-rooted organization patterns, which make collective social behavior comply with the fundamental laws governing the past, the present, and the future.

INVARIANTS

The simplest possible law dictates that something does not change—an *invariant,* in scientific terms. Invariants are, of course, the easiest things to forecast. They reflect states of equilibrium maintained by natural regu-

---

is the bringing of fingerprinting to Scotland Yard. He also wrote an essay pointing out that the number of brush strokes required for a portrait is about 20,000, very close to the number of hand movements that go into the knitting of a pair of socks.[3]

lating mechanisms. In ecosystems such equilibrium is called *homeostasis* and refers to the harmonious coexistence of predator and prey in a world where species rarely become extinct for natural reasons.

States of equilibrium can also be found in many aspects of social living. Whenever the level of a hardship or a menace increases beyond the tolerable threshold, corrective mechanisms are automatically triggered to lower it. On the other hand, if the threat accidentally falls below the tolerated level, society becomes blasé about the issue, and the corresponding indicators start creeping up again with time.

Invariants have the tendency to hide behind headlines. For example, the number of deaths due to motor vehicle accidents becomes alarming when reported for a big country like the United States over a three-day weekend, which is what journalists do. However, when averaged over a year and divided by one hundred thousand of population, it becomes so stable over time and geography that it emerges as rather reassuring!

Safety in Cars

Car safety is a subject of great interest and emotion. Cars have been compared to murder weapons. Each year close to two hundred thousand people worldwide die from car accidents, and up to ten times as many may suffer injuries. Efforts are continually made to render cars safer and drivers more cautious. Have such efforts been effective in lowering the death rate? Can this rate be significantly reduced as we move toward a more advanced society?

To answer those questions we must look at the history of car accidents, but in order to search for a fundamental law we must have *accurate* data and a *relevant* indicator. Deaths are better recorded and more easily interpreted than less serious accidents. Moreover, the car as a public menace is a threat to society, which may "feel" the dangers and try to keep them under control. Consequently, the number of deaths per one hundred thousand inhabitants per year becomes a better indicator than accidents per mile or per car or per hour of driving.

The data shown in Figure 1.1 are for the United States starting at the beginning of the century.[5] What we observe is that deaths caused by car accidents grew with the appearance of cars until the mid-1920s, when they reached about twenty-four per one hundred thousand per year. From then onward they seem to have stabilized, even though the number of cars continued to grow. A homeostatic mechanism seems to enter into action when this limit is reached, resulting in an oscillating pattern around the equilibrium position. The peaks may have produced public outcries

SAFETY IN CARS

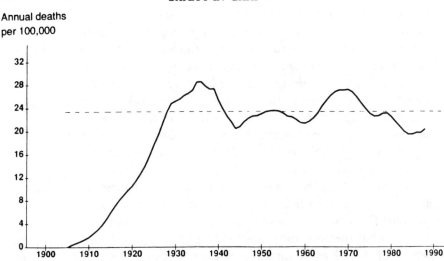

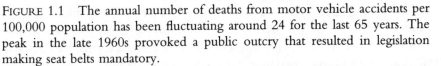

FIGURE 1.1 The annual number of deaths from motor vehicle accidents per 100,000 population has been fluctuating around 24 for the last 65 years. The peak in the late 1960s provoked a public outcry that resulted in legislation making seat belts mandatory.

for safety, while the valleys could have contributed to the relaxation of speed limits and safety regulations. What is remarkable is that for sixty-five years there has been a persistent self-regulation on car safety in spite of tremendous variations in car numbers and performance, speed limits, safety technology, driving legislation, and education.

Why the number of deaths is twenty-four per one hundred thousand per year, and how society can detect a rise to twenty-eight, is not clear. One can look at other countries for possible cultural links. For the year 1975 seven nations (Austria, Belgium, France, Denmark, Italy, Canada, and the United States) had death rates of around twenty-four. Three countries were exceptions. For the United Kingdom, Sweden, and Japan the number was around twelve. All three countries were driving on the left at that time. If this difference is significant, exploring this phenomenon further may be interesting. Could it be that left-hand drivers are less accident-prone?

The number of deaths in the last fifteen years suggests a downturn. Are we finally improving in car safety, or is it yet another fluctuation? American society has been tolerating this level for over sixty years. A Rand analyst has put it rather bluntly: "I am sure that there is, in effect, a desirable level of automobile accidents—desirable, that is, from a broad point of view, in the sense that it is a necessary concomitant of things of greater value to society."[6] Lacking compelling arguments to the contrary, my best bet for a long-term forecast would be little or no change, in what appears to be an invariant related to car safety. In fact, given that in recent years the number of deaths per one hundred thousand per year has been closer to twenty than to the "canonical" twenty-four, I would expect a rise of about 20 percent during the decade of the 1990s. Such a rise will probably be explained later as having been caused by the ever-increasing number of cars on the road as well as by the upward drifting of the speed limits from their present low settings.

An invariant can be thought of as a state of well-being. It has its roots in nature, which develops ways of maintaining it. Individuals may come forward from time to time as advocates of an apparently well-justified cause. What they do not suspect is that they may be acting as unwitting agents of deeply rooted necessities for maintaining the existing balance, which would have been maintained in any case. An example is Ralph Nader's crusade for car safety, *Unsafe at Any Speed,* published in the 1960s, by which time the number of fatal car accidents had already demonstrated a forty-year-long period of relative stability. But examining Figure 1.1 more closely, we see that the 1960s show a small peak in accidents, which must have been what prompted Nader to blow the whistle. Had he not done it, someone else would have. Alternatively, a timely social mechanism might have produced the same result; for example, an "accidental" discovery of an effective new car-safety feature.

Rationing Computer Innovation

An invariant surfaced accidentally in the world of computers while I was studying technological innovations together with my colleague Alain Debecker. We were investigating the rate of appearance of new computer models along with the emergence of new computer manufacturers within the same time frame. Our impulse was to relate models to manufacturers, so we checked how the number of new models depended on the number of new manufacturers. We were pleasantly not surprised to witness a perfectly straight line being traced under our eyes, dot after dot, on the computer screen. During the first 25 years of computer history and over the full range of computer models, there appears to have existed an invariant, a hidden regulation carefully respected. Each manufacturer seems to be entitled to only a small number of computer models, five on the average.

We stared at this for a while, unable to reconcile it with large companies like IBM and DEC, which are credited with hundreds of models. In the end it became clear. In these gigantic citadels of computer know-how, successful models, concepts, and individuals branch off to give rise to small companies, often based on a single idea or person. Major computer manufacturers do not only create machines, they also create new companies at the strictly obeyed ratio of five to one (more about this in Chapter Four.)

Humans Versus Machines

Invariants characterize a natural equilibrium. The average human sleeps during one third of the day-night cycle. Similarly, the optimum amount of time spent traveling daily is seventy minutes. It seems reasonable, then, that there should also be a balance between the amount of physical work and the amount of intellectual work carried out during a day. In a loose comparison of computers to living organisms, software can be likened to intellect and hardware to body, and there appears to be an invariant emerging in the world of computers that contradicts the best judgment of some information technology experts.

People feel insulted when they are compared to animals. Yet, strangely, they take pride in themselves when they do something as well as a machine, perhaps because they believe this resemblance cannot be taken seriously. From the discovery of the wheel to the discovery of the transistor, history is punctuated with milestones marking the appearance of machines that relieved humans from repetitive burdens. Industrialization featured mostly muscle-surrogate inventions, but that did not significantly decrease the number of working hours. Allowing eight hours for sleep and a fair amount for personal matters, the time available for work cannot be far from eight to ten hours per day. At the same time, human nature is such that a much shorter work period is poorly tolerated.

Soon after the introduction of computers, the need for software gave rise to a thriving new industry, because a computer is useless until programmed to do a task. The fast growth of successful software companies triggered speculation among computer manufacturers that someday computers might be given away free; all revenue would be made from services and software. This meant that the computer industry's major effort would ultimately become the production of software.

Such a trend actually began to develop. In the 1970s large research institutions built computer-support departments heavily staffed with programmers. Hordes of them devoted their youth and talent to writing thousands of lines of FORTRAN to offer scientists the possibility of making graphs, histograms, and computer-generated drawings. That situation soon became intolerable, though not necessarily consciously, and graphic capabilities were progressively transferred to the hardware. Today video terminals provide artwork without even bothering the central processing unit of the computer with such menial tasks. Meanwhile, programmers have moved on to a higher challenge. They spend their time increasing manyfold a computer's performance through parallel processing techniques, which themselves will soon become fully incorporated into the hardware.

For at least eight years now the hardware-to-software ratio of expenditures has been stable. In Western society the natural equilibrium level shows this ratio to be about three to one. This fact can be seen as a reflection of the homeostasis between human work (software) and machine work (hardware) in the information technology industry.

While invariants reveal the average long-term trend, there are fluctuations above and below the natural level. Such fluctuations can occasionally reach significant dimensions and play an important role in society. In the realm of human versus machine work, there are few social situations in which excessively mechanical behavior is tolerated, condoned, or tacitly encouraged. One example is people performing on musical instruments. Technical virtuosity on any musical instrument is cherished; it accounts for most of the effort but never wins the highest praise. A typical unfavorable critique of a performance may revolve around such phrases as "admirable technique, but. . . ." There seems to be a reticence on the part of both performers and critics to praise technical dexterity in proportion to the effort that has gone into it. This is not surprising, considering that it is indeed largely mechanical behavior. But then, why do they all vehemently oppose having a machine actually do the mechanical part of the work for them? The following story is a thought experiment involving a hypothetical machine called the Aponi.[7]

. . .

We are in Philharmonic Hall at Lincoln Center in 1999 for an unusual concert. The New York Philharmonic Orchestra, seated on the stage, has just finished tuning its instruments for a performance of Mozart's piano concerto No. 21. Conductor and soloist, both dressed in black tails, come onto the stage as the audience applauds. The soloist is carrying a long rectangular black box—the Aponi—under his arm. He carefully places it on the Steinway, thus covering the keyboard, and plugs a cable extending from one end of the box into an outlet on the floor. He sits on the pianist's stool and places his hands on top of the Aponi. He slides his fingers into three continuously variable ergonomic switches.

The Aponi, an interface to the keyboard, is equipped with little hammers driven by electromagnets and a diskette-reading unit. The music score for the piano part has already been optically read and recorded on the diskette. The Aponi is capable of "playing" this score flawlessly and at any speed. Furthermore, it offers the possibility of personalizing the interpretation (via the switches) in the only three possible ways: faster/slower, piano/forte, and staccato/legato. All mechanical work is done by the machine. All intelligent work—the interpretation—is left to the soloist who can express his or her inner self fully while performing the most difficult piano pieces effortlessly.

Pianist and conductor exchange a glance, and the concert begins. In a little while the familiar soothing melodies dampen the agitation of the audience. People become less aware of the intruding machine as they notice the usual movements of the soloist, leaning forward and backward—to play forte he has to push down on the finger switches. But at the end of the concert the agitation builds again, particularly as rumors start circulating that the soloist is no musician but a second-generation Italian restaurant owner who likes opera and classical music.

Even more anger is expressed by critics in the following day's press. Musicians, and particularly pianists, feel the rug has been pulled from under their feet. Men and women who have spent their youth laboring over keyboards, sometimes damaging their bodies in their quest to acquire superhuman techniques, bitterly resent this competition from a mere amateur.

The most annoying consequence of the new invention is the embarrassment it causes among all lovers of live music. Appreciation in concert halls is purportedly directed toward the performer's gift to interpret. His or her technical ability always comes in second place. Now that a machine has made it possible for anyone to interpret without effort (the Aponi does all the mechanical work), there is not enough dynamic range in interpretation for rating all the good performers. It becomes clear, suddenly, why capable pianists choose difficult pieces for their programs (or play easier pieces impossibly fast): to distinguish themselves by their exceptional technical virtuosity and thus become eligible for a judgment on interpretation. Technical virtuosity has become a means of eliminating the large numbers of good pianists for whom there is not enough room on the stages of concert halls. A disturbing question, however, is whether the individuals capable of acquiring a "freaky" manual dexterity are also gifted with the artistic beauty that is expected from their interpretation.

• • •

Performances on musical instruments are examples of social situations in which human work and mechanical work are far from being reasonably balanced. Trying to keep alive an older musical tradition (instruments, compositions, performers, standards of criticism, and so forth) in today's overpopulated world of high competition and people difficult to impress has probably accentuated this imbalance. Social living has produced performers who stand as exceptions to a natural equilibrium. There is an invariant in the animal kingdom, however, where civilization has caused little or no deviation from the norm.

Do All Animals Die at the Same Age?

The notion that all animals die at the same age sounds implausible if you measure age in years and months, but it becomes rather logical if you count the number of heartbeats. The Greek professor of preventive medicine Vasilios Valaoras claims that most mammals living free in nature (not in homes and zoos) have accumulated about one billion heartbeats on the average when they die.[8] It is only the rate of the heartbeat that differs from animal to animal. Small ones, like mice, live about three years, but their heartbeat is very rapid. Middle-size ones, rabbits, dogs, sheep, and so forth, have a slower heartbeat and live between twelve and twenty years. Elephants live more than fifty years but have a slow heartbeat. Even before Valaoras, Isaac Asimov had remarked, "Whatever the size . . . the mammalian heart seems to be good for a billion beats and no more."[9]

For hundreds of thousands of years humans had a life expectancy between twenty-five and thirty years. With the normal rate of seventy-two heartbeats per minute, they conformed nicely to the one billion invariant. Only during the last few hundred years has human life expectancy significantly surpassed this number, largely due to reduced rates in infant mortality from improved medical care and living conditions. To-

day's humans reach three times the mammals' quota, positioning Homo
sapiens well above the animal world. Are we finally making progress in
our evolution as a species?

"On the contrary, we are regressing!" exclaims Marchetti in his usual
playing-the-Devil's-advocate style. Life expectancy at birth increased primarily because infant mortality decreased. But what also increased at the
same time was the availability and acceptability of safe and legal abortions,
resulting in a rise of *prenatal* mortality, thus canceling a fair amount of the
life expectancy gains. The end result, he says, claiming to have seen the
data that substantiate it, is that life expectancy *at conception* is still not much
above forty.

If there is any truth in this, we are back—or close enough—to the
one-billion-heartbeat invariant, but with an important difference. Low
infant mortality rates result in the birth of many individuals who may be
ill-suited to survive a natural selection process favoring the fittest. At the
same time, abortions are blind. They eliminate lives with no respect to
their chance of survival. A selection *at random* is no selection at all, and the
overall effect for the species is a degrading one.

With life expectancy defined at conception, the one-billion-heartbeat
invariant seems to still be roughly in effect for humans. From now onward there can be no significant further gains in infant mortality rates
because they are already quite low. Moreover, as we will see in Chapter
Five, the overall death rate has practically stopped decreasing. All this
argues for bleak forecasts about life expectancy growth. Valaoras believes
that there is a message for us coded in the one-billion-heartbeat invariant.
His thesis is that we are now reaching the upper limit in life expectancy,
and he cites the Bible as a much earlier statement of this conclusion:[10]

> Seventy years is all we have—eighty years, if we are strong; yet all they
> bring us is trouble and sorrow; life is soon over, and we are gone. (Psalm
> 90:10).

On the other hand, an invariant turning into a variable is nothing
unusual. A frequent discovery in physics is that something thought of as
a constant is, in fact, a limited manifestation of a simple law, which in turn
is a special case of a more complicated law, and so on. When the complexity becomes unmanageable, physicists turn toward unification,
searching for a general law that will describe a multitude of diverse
phenomena, each hitherto abiding by elaborate custom-made rules. In
other words, they look for a new invariant on a higher abstraction level.

For example, the acceleration of gravity, 10m/sec/sec, had been perceived as a constant from the early falling-apple days until Newton showed that the rate decreases with altitude. Later Einstein's relativity theory revealed more complicated equations for the falling apple. Today, besides apples, there are hundreds of elementary particles, all behaving according to rules made to measure. The physicists' dream in the meantime has become the Grand Unification wherein the behavior of apples as well as particles, stars, galaxies, and light can be described by the same set of equations.

To predict the future we must first find out the constraints that have to be met, the invariants and other natural laws that have influenced or governed in the past and will, in the highest degree of probability, continue to do so in the future. There are many mathematical equations that can describe such certainties, but I have found that much of their essence is captured visually through the shapes associated with these equations, which accurately depict the past and provide reliable forecasts for the future.

THE BELL-SHAPED CURVE OF THE LIFE CYCLE

Invariants are the simplest possible natural laws that exist for all time, without a beginning, middle, or end, and which could be visualized as a straight line. However, there are natural laws with increasing complexity but with invariant overall form having a beginning and an end. Such is the law of natural growth over a span of time wherein a beginning is linked with growth and a decline with death. Coming into and out of existence is often called a *life cycle,* and its popular pictorial symbol is a bell-shaped curve shown in Figure 1.2.

Originally used in biology, the bell curve illustrates that anything with life grows at a rate which goes over a peak halfway through the growth process. In the case of a child's height, for example, the annual growth rate reaches a maximum around nine years of age. If Figure 1.2 were to represent the annual rate of growth, it would have an approximate time scale in years, so that "Birth" appears at zero years, "Growth" at around four to five years, "Maturity" around nine, "Decline" at thirteen to fourteen, and "Death" at eighteen years of age, beyond which no more growth is expected. If, on the other hand, Figure 1.2 were to represent the "life cycle" of a pianist's career—expressed in terms of the number of concerts given annually—an approximate time scale would indicate

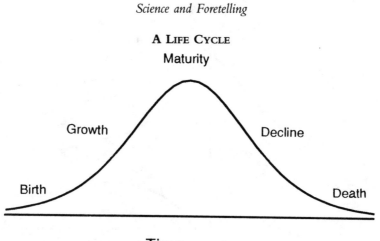

FIGURE 1.2   This bell-shaped curve is used across disciplines as a template for the cycle of life.

"Birth" at around twenty years of age, "Growth" at thirty, "Maturity" at forty-five, "Decline" at sixty, and "Death" around seventy. The bell-shaped curve would qualify the pianist as a promising young performer at twenty, gaining popularity at thirty, world renowned at forty-five, and in decline at sixty.

The concept of a life cycle has been borrowed by psychologists, historians, businessmen, and others to describe the growth of a process, a product, a company, a country, a planet. In all cases it represents the *rate* of growth, which is zero before the beginning and becomes zero again at the end. The bell-shaped curve of Figure 1.2 has become a visual symbol for natural growth and serves most often only qualitatively. It can also be of quantitative use, however.

THE S-CURVE OF CUMULATIVE GROWTH

As one moves along the life cycle curve passing through the various stages, the cumulative number of the units represented by the bell curve grows in such a way that it traces the shape of an S, shown at the lower part of Figure 1.3. The S-curve can therefore become a visual symbol for cumulative growth. For example, while the *rate* of growth of a child's height traces a bell-shaped curve, the *overall* size of the child produces an S-curve going roughly from two and a half feet at birth, to six feet, eighteen years later. Halfway through, when the growth rate is fastest, the

NATURAL GROWTH

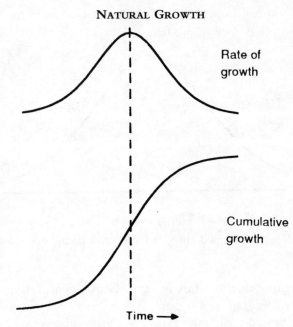

FIGURE 1.3   While the rate of growth follows a bell-shaped curve, cumulative growth traces an S-curve.

child's height is around four feet three inches, that is, the original two feet six inches plus one foot nine inches, which is half of whatever he or she will ever gain through adolescence (prenatal growth is a separate process not considered here). These are average heights, of course, and there can be substantial variation from one individual to another.

As for the other example mentioned earlier, the career of a pianist, the rate at which he gives concerts over the lifetime of his career traces a bell-shaped curve, but the cumulative number of concerts he has given up to a certain time traces an S-curve.

A last example is the launching of a new product. The life cycle of the product can be measured by the number of units sold per month and follows a bell-shaped curve, while the cumulative number of units sold until any given month follows an S-curve.

The overall size obtained as a cumulative number, be it height, number of concerts, or products sold, can never decrease, although in real life a person's height does, in fact, decrease somewhat with old age.

Just as an S-curve can be obtained from a life cycle by forming the cumulative rate of growth over time, one can calculate a life cycle from the S-curve through successive subtractions. For example, monitoring a child's growth over the years by plotting his or her height on graph paper produces an overall S-curve. If one is interested in the life cycle of the process, one must subtract from each year's measurement the previous year's, thus producing a set of numbers representing growth per year, which will follow the bell-shaped curve of the natural life cycle.

THE LAW OF NATURAL GROWTH

The law describing natural growth has been put into mathematical equations called growth functions. The simplest mathematical function that produces an S-curve is called a *logistic* and is described in detail in Appendix A. It is derived from the law which states that the rate of growth is proportional to both the amount of growth *already accomplished* and the amount of growth *remaining to be accomplished*. If either one of these quantities is small, the rate of growth will be small. This is the case at the beginning and at the end of the process. The rate is greatest in the middle, where both the growth accomplished and the growth remaining are sizable. Furthermore, growth remaining to be accomplished implies a limit, a saturation level, a finite market size. This ceiling of growth is assumed to be constant throughout the growth process. Such an assumption is a good approximation of many natural-growth processes, for example, human growth, in which the final height is genetically precoded.

It is a remarkably simple and fundamental law. It has been used by biologists to describe the growth under competition of a species population, for example, the number of rabbits in a fenced-off grass field. It has also been used in medicine to describe the diffusion of epidemic diseases. J. C. Fisher and R. H. Pry refer to the logistic function as a *diffusion model* and use it to quantify the spreading of new technologies into society.[11] One can immediately see how ideas or rumors may spread according to this law. Whether it is ideas, rumors, technologies, or diseases, the rate of new occurrences will always be proportional to how many people have it and to how many people don't have it yet.

The analogy has also been pushed to include the competitive growth of inanimate populations such as the sales of a new successful product. In the early phases, sales go up in proportion to the number of units already sold. As the word spreads, each unit sold brings in more new customers. Sales grow exponentially. It is this early exponential growth that gives rise to the first bend of the S-curve. Business looks good. Growth is the same percentage every year, and hasty planners prepare their budgets that way. Growth, however, cannot continue to be exponential. Explosions are exponential. Growth follows S-curves. It is proportional to the amount of the market niche *still unfilled*. As the niche fills up, growth slows down and goes into the second bend of the S-curve, the flattening out. Finally we reach zero growth and the end of the life cycle; the growth process in question comes to an end. The bell curve depicting the rate of natural growth goes back to zero, while the S-curve of cumulative growth reaches its ceiling.

What is hidden under the graceful shape of the S-curve is the fact that natural growth obeys a strict law which is seeded with knowledge of the final ceiling, the amount of growth *remaining to be accomplished*. Therefore, accurate measurements of the growth process, even during its early phases, can be used to determine the law quantitatively, thus revealing the final size (the value of the ceiling) ahead of time. This is why the S-curve approach possesses predictive power.

Controls on Rabbit Populations

There is an inherent element of competition in logistic growth—competition for space, food, or any resource that may be the limiting factor. This competition is responsible for the final slowing down of growth. A species population grows through natural multiplication. Besides feeding, rabbits do little else but multiply. The food on their range is limited, however, with a capacity for feeding only a certain number of rabbits. As the population approaches this number, the growth rate has to slow down. How does this happen? Perhaps by means of increased kit mortality, diseases, lethal fights among overcrowded rabbits, or even other more subtle forms of behavior that rabbits may act out unsuspectingly. Nature imposes population controls as needed, and in a competitive environment, only the fittest survive.

Since the growth of a rabbit population follows an S-curve from the beginning, one may conclude that these natural controls are so sophisticated that they take effect from the very first rabbit couple on the range. The same mathematical function describes the process from beginning to end. The function can be determined, in principle, after only a few early measurements on the rabbit population. Theoretically, therefore, as soon as this function is defined, the fate of the growth of the rabbit population is sealed.

I became suspicious the first time I was confronted with this notion. I could not explain how the first rabbit couple would have to behave "in accordance" with a final limitation it had not yet felt. My doubts were resolved when I realized that the early part of an S-curve is practically indistinguishable from an exponential. In fact, the rabbits behave as if there are no controls. If the average rabbit litter is taken as two, then we observe the rabbit population go through the successive stages of 2, 4, 8, 16, 32, 64, . . ., 2^n in an exponential growth. There is a population explosion up to the time when a sizable part of the niche is occupied. It is only after this time that limited food resources start imposing constraints on the number of surviving rabbits. That is why trying to determine the final ceiling from very early measurements may produce enormous errors.

In addition, all measurements are subject to fluctuations. Special conditions such as bad weather or a full moon may perturb the reproductive process. Similarly for the analogous case of product sales, the numbers are subject to political events, truck driver strikes, or stock market plunges. Small fluctuations on a few early measurements, if that is all we use, can make a huge difference on the estimate of the final niche capacity.

That is why the S-curve can be meaningfully determined only if the growth process has proceeded sufficiently to deviate from the early exponential pattern. It is advisable to determine the curve from *many* measurements and not just from a few early ones. This way we diminish the impact of rare events, and measurement errors will largely average out. The correct procedure is a *fit* described in detail in Appendix A. A computer is programmed to search through trial and error for the S-curve that passes as close as possible to as many data points as possible. The curve chosen will provide us with information outside the data range, and in particular with the value and the time of onset of the final ceiling.

Critics of S-curves have always raised the question of uncertainties as the most serious argument against forecasts of this kind. Obviously the more good-quality measurements available, the more reliable the determination of the final ceiling. But I would not dismiss the method for fear of making a mistake. Alain Debecker and I carried out a systematic study of the uncertainties to be expected as a function of the number of data points, their measurement errors, and how much of the S-curve they cover. We did this in the physics tradition through a computer program simulating "historical" data along S-curves smeared with random deviations, and covering a variety of conditions. The subsequent fits aimed to recover the original S-curve. We left the computer running over the weekend. On Monday morning we had piles of printout containing over forty thousand fits. The results were published,[12] and a summary of them is given in Appendix B, but the rule of thumb is as follows:

> If the measurements cover about half of the life cycle of the growth process in question and the error per point is not bigger than 10 percent, nine times out of ten the final niche capacity will turn out to be less than 20 percent away from the forecasted one.

The results of our study were both demystifying and reassuring. The predictive power of S-curves is neither magical nor worthless. Bringing this approach to industry could be of great use in estimating the remaining market niche of well-established products within quoted uncertainties. Needless to say life is not that simple. If products are not sufficiently differentiated, they are sharing the niche with others, in which case the combined populations must be considered. Furthermore, the fitting procedure as described in Appendix B allows for weights to be assigned to the different data points. The wise user can thus put more emphasis on some points than on others. The end result is value added by the person who is doing the fit. This methodology does not automatically produce identical results by all users. Like any powerful tool, it can create marvels in the hands of the knowledgeable, but it may prove deceptive to the inexperienced.

GAUSS FOR STATISTICS WAS LIKE BACH FOR MUSIC—AN INGENIOUS FORGER

A logistic life cycle is remarkably similar to the bell-shaped distribution curve usually referred to as "normal" (Appendix C, Figure 1.1). The curve is called Gaussian* in textbook literature and has been overused by hard and soft sciences to describe all kinds of distributions. Examples range from people's intelligence (IQ index), height, weight, and so forth, to the velocities of molecules in a gas, and the ratio of red to black outcomes at casino roulette wheels.

All this glory to Mr. Gauss, however, is somewhat circumstantial. A well-known man of science, he was referred to as "the prince of math-

---

* Karl Friederich Gauss (1777–1855) was a German mathematician who in the nineteenth century publicized a bell-shaped curve now called the normal distribution function. The first encounter of this function in the mathematical literature, however, is attributed to the French mathematician Abraham de Moivre (1667–1754).

ematics" in nineteenth-century literature, and the bell-shaped curve he provided resembled very much the distributions with which people were concerned. In addition, it is obtained through an elegant mathematical derivation. For historical reasons mostly, people proceeded to put Gaussian labels on most of the variables on which they could get their measuring sticks. In reality there are not many phenomena that obey a Gaussian law precisely.

The logistic life cycle is so close to a Gaussian curve that they could easily be interchanged. No psychologist, statistician, or even physicist would ever have come to a different conclusion had he or she replaced one for the other in a statistical analysis. What is ironic is that the Gaussian distribution is called "normal" while there is no natural law behind it. It should rather have been called "mathematical" and the name "normal" reserved for the S-curve law that is so fundamental in its principles.

This situation echoes something that happened in music. It was for convenience sake that Bach introduced the well-tempered way of tuning musical instruments. A mathematical genius, Bach was drawn to compose on all possible keys. The twelve semitones with the options of major or minor offer twenty-four different keys for composition. Up to that time less than a handful of them were being used by composers. The other keys produced dissonance for the normally harmonious chords of thirds and fifths.

Bach succeeded in finding a way to tune his harpsichords that allowed him to compose in all twenty-four keys. The proof lies with his celebrated work *The Well-Tempered Clavier.* The century that followed saw the widespread popularity of pianos as well as the appearance of brilliant mathematicians. Pianos made long-lasting tuning imperative, and mathematicians elegantly formulated Bach's way of tuning as setting the ratio of frequencies between successive semitones equal to $\sqrt[12]{2}$ (the twelfth root of 2). In so doing they produced *equal-temper* tuning, pushing Bach's idea one step further. (Contemporary musicologists argue that Bach may have disapproved of this!) Thus transposing became possible, musical ensembles flexible, and life generally easier for all musicians.

As it happens, equal-temper tuning is very close to the natural octave so that the differences are almost imperceptible. The natural octave, however, may have more profound reasons for existence. Purist musicians claim that piano and violin can no longer play together and sometimes demand an old-fashioned tuning for a favorite recital. Furthermore, studying the natural octave may provide insights and understanding on a cosmological level. In his book, *In Search of the Miraculous,* Peter Ouspensky, the Russian mathematician and philosopher, recounts ideas that relate the natural octave to the creation of the universe![13] The seeker of pure and esoteric truth should be wary of elegant formulations such as those of Bach and Gauss.

Independently of arguments of elegance, the logistic function is intimately associated with the law of natural growth. One can visualize this law either with the bell-shaped or with the S-shaped curve. Both are seeded with a parameter describing the ceiling, the capacity of the niche in the process of being filled. The important difference between them, from the statistical analysis point of view, is that the S-curve, which usually depicts a cumulative rate of growth, is much less sensitive to fluctuations because one year's low is compensated for by another year's high. Therefore, the S-curve representation of a natural growth process provides a more reliable way to forecast the level of the ceiling. From an intuitive point of view, an S-curve promises the amount of growth that can be accomplished while a bell-curve heralds the coming end of the process as a whole.

2

Needles in a Haystack

HOW MUCH ORDER IS THERE IN "RANDOM" DISCOVERIES?

LUCY: What are you doing on top of that haystack?

CHARLIE BROWN: I'm looking for needles.

LUCY: You'll never find one.

CHARLIE BROWN: Sure I will. There is a lot of them in here.

LUCY: How do you know?

CHARLIE BROWN: I threw them in so I can practice finding them.

LUCY: It will take you forever.

CHARLIE BROWN: It may be slow in the beginning, but as I get better, it will go faster and faster. The more needles I find, the faster it will go.

LUCY (stubbornly, walking away): It'll take you forever finding the last one!

• • •

Lucy and Charlie Brown are both right. They are arguing simply because each one is looking at a different aspect of learning.

Learning is a process that has a beginning, a phase of growth, maturity, decline, and an end like the life cycle of any other natural growth process. From academia to industry the process of learning has been associated with a curve that is markedly similar to the S-curve described in the previous chapter. Learning starts slowly, proceeds rapidly during the middle of the process (the steep rise of the S-curve representing the cumulative amount of learning achieved), and slows down sometime later. At the end, and as the S-curve reaches its ceiling, the rate of learning ceases altogether. Learning curves are usually used only in a qualitative way. The intriguing question is whether learning is governed by the law of natural growth under competition, in which case it should obey the logistic mathematical function *quantitatively* also.

During the early phases of a certain activity, repetition accelerates learning. The more you do something, the faster you can do it the next time around. The rate of increase in learning seems to be more rapid than simply proportional to the number of repetitions. Imagine that you are sitting in front of a typewriter for the first time. Even though all the letters are clearly labeled, it will take a while before you can type your name. The third time you type your name, however, you will probably do it more than three times faster than the first time. This early exponential growth of learning corresponds to the first bend of the S-curve. Soon, however, you reach a point where the rate of improvement in speed slows down. You are exhausting the capabilities of your muscle reactions. You may want to improve on that, but it will be a slow, painful process with diminishing returns.

In other forms of learning you may find that you are exhausting the amount of learning that can be achieved *before* you reach the limit of your capabilities. For example, you can improve your performance at playing tic-tac-toe through repetitions up to the point at which you have mastered the game. From then onward, further improvement becomes impossible, no matter how much you practice. Not so with chess; here the number of configurations is so vast that they can be considered inexhaustible. What becomes exhausted is your ability to retain an increasingly larger number of configurations in your mind. You may also become exhausted physically or even exhaust the time available for playing chess. In all cases learning will slow down and the final level of achievement will vary from one individual to another.

The slowing down of the learning process as one approaches the saturation level results in the second bend of the S-curve. The ultimate reason for this slowdown is some form of competition, be it for the use of one's time, energy, or physical capabilities. The rate at which you are able to find needles in a haystack declines because there are fewer needles to be found. The proof, however, of whether learning processes obey the natural-growth law quantitatively or not rests with the actual data. Let us look closely at some typical learning processes.

As a first example, consider the evolution of an infant's vocabulary. Acquiring vocabulary is a random search similar to finding needles in a haystack. The "needles" to be found are the words in the combined active vocabulary of the two parents. The words most frequently used will be learned first, but the rate of learning will eventually slow down because there are fewer words left to learn. In Figure 2.1 we see the evolution of an infant's vocabulary during the first six years.[1]

The best-fitting S-curve follows the data points closely to reach a plateau of twenty-five hundred words by the end of the sixth year. This ceiling defines the size of the home vocabulary "niche," all the words available at home. Later, of course, schooling enriches the child's vocabulary, but this is a new process, starting another cycle, following probably a similar type of curve to reach a higher plateau.

This pattern of learning is not restricted to individuals. It is also encountered with a well-defined group of people, a country or humanity as

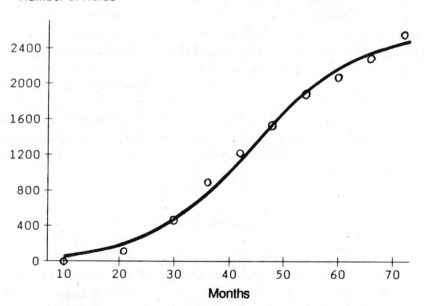

CHILDREN LEARN ACCORDING TO LEARNING CURVES

Number of words

FIGURE 2.1 The S-curve fitted to the data points shows the way infants acquire vocabulary until they exhaust the home "niche," typically limited to about 2.500 words.

a whole. For example, the scientific community has followed this same pattern, as if it were a single individual, while discovering the stable chemical elements throughout the centuries. If we make a graph of the number of elements known at any given time, an accumulation of knowledge, we obtain a pattern that resembles the S-curve of an infant's learning process. The main difference is that the time scale is in centuries instead of years. What may have been thought of as random discoveries are seen to be associated with a regular curve.

The first dozen elements had been known for a very long time; iron, copper, gold, mercury, and carbon were familiar to the men of antiquity. In the middle of the eighteenth century, however, there was a thirst and a push for discovering more and more stable elements, and by the mid-twentieth century all the naturally available elements had been found. Still, researchers pushed for more discoveries, now carried out in particle accelerators. These late entries—one may say concoctions—hardly deserve the name "stable" elements because they last only a short time before they decay. At any rate, we have been running out of those, too. The last artificially created element was established in 1960. Overall, the 250-year-long discovery process seems to have followed the path of natural growth and reached completion.

In both learning processes discussed above, we see "random" discoveries following overall orderly curves, an apparent contradiction since randomness is associated with chaotic behavior while S-curves represent the order imposed by the logistic law. It can be shown, however, that chaos is seeded with S-curves to begin with, and James Gleick in his book *Chaos* explains how chaos can result from a prolonged action of the logistic law. One can find an "eerie sense of order" in chaotic patterns. We will be looking more closely at the relationship between order and chaos in Chapter Ten, and will conclude that such a type of order should not be so eerie after all.

LEARNING FROM SUCCESS

Collective learning patterns can also be found in a group of people with a common goal of manufacturing and selling products in order to make a profit; in other words, a company. The assumption is that companies learn like individuals, that is, according to learning curves. The objective is to learn how to make and sell products successfully. For technological products part of the learning is in mastering the technology, and another part is how to reduce the costs of production. But success in business is not solely a question of technology and price. A product will not sell well unless it matches the needs of the customer who buys it.

Research engineers are often geniuses and may produce machines that are too good. A product offering features that people are not yet ready to exploit is likely to sell poorly. One remedy is to keep such a product in storage and launch it later when the time is right. Another approach is to raise its price for the few who can use it and hope to make a profit in this way despite sparse sales. In both cases the crucial issue is to relate performance to price in a way that is optimum at the time. The relationship between performance and price is a subject that can be studied and learned. Historical records contain information on how this relationship evolved over time and how it affected the sales of successful products. As in natural selection, success is keyed to survival of the fittest. Competitive growth is known to follow S-shaped patterns, and so does learning. Thus, the elements are all there for tracing a learning curve for success or, more precisely, the continuation of an already established successful process.

Could You Take Aim at an Apple on Your Son's Head?

The world of business is an arena where ambitious and motivated individuals compete fiercely in order to make a profit. Competitive advantages are sought by any means, generally complying with society's laws but always honoring nature's laws. As in the jungle, success—sometimes equivalent to survival—favors the fittest.

The fittest is the one with the most competitive advantages. In times of famine the fittest may also be the fattest, which translates to business mottoes like "bigness is goodness." At other times, however, the fittest is the one who does something new and different, especially if it turns out later that this new direction was a hidden gold mine.

Successful business persons use their instincts to "smell out" profitable directions. Instead of systematic methodologies, they use their intuition— "my tummy tells me." Those who are gifted that way succeed more often than not. Those who turn to educational institutions for their competitive advantage have been sparingly rewarded. The know-how transmitted through a master's in business administration has rarely proved an effective weapon against a gifted tummy. Unlike renowned professionals in the arts and the sciences, successful entrepreneurs have not usually given much credit to their university professors. Most often they make references to some popular, folkloric, street-wise education gained outside schools or to some genetic heritage.

Instinctive performance can and has gone a long way, particularly during prosperous times and conditions that permit one to "dance" one's way through good business decisions. The difficulties show up during economic recessions when markets start saturating, competition becomes cutthroat, and the name of the game changes from prosperity to survival. As in the performances of musicians or acrobats, heavy psychological stress results in insecurity, panic, and mistakes. William Tell may have been the rare exception, daring to take aim at the apple on his son's head. Most good archers would have hands shaking too much under the stress of what was at stake.

Stress was quite evident among the executives who said to me, "We had no difficulty positioning our products back when business was booming, but now that it has become of crucial importance, how can we be sure we are not making a mistake?" They were simply expressing their need for systematic guidelines in order to continue doing what they had been doing, apparently by instinct. Are there such guidelines? I believe there are. By quantifying the learning from past successful decisions, marketing executives can devise a systematic way to position future products.

A Real Case

The performance of a computer is gauged by how many **m**illions of **i**nstructions it can execute **p**er **s**econd (MIPS). The figure of merit for a computer model then becomes the ratio performance/price expressed in MIPS/$. I studied Digital's VAX family products from the very first model, the VAX 11/780, to the much more recent mainframes. Like those of other companies, Digital's research engineers have been creating increasingly powerful computers for less money. The ratio MIPS/$ for all VAXes shows them to be following S-shaped learning curves.[2] The learning process is still young because in general the data occupy only the early exponential-like part of the curve. Consequently, estimates of the final ceiling carry large uncertainties (of up to 100 percent). Our a priori knowledge of a level for this ceiling is so poor, however, that even very uncertain estimates are useful.

It is interesting to know, for example, that we have a long way to go. This may sound like a trivial conclusion, but thanks to the way it was obtained, it says more than simply "man will always be making progress."

It says that a certain amount of computer evolution is guaranteed to us by the fundamental technologies available today. Breakthroughs will certainly come along. Nothing as dramatic as the discovery of the transistor, however, will be warranted in order to reach three to four times today's levels of performance/price. Expectedly, the appearance of an earth-shaking breakthrough, a discovery of importance unprecedented during computer history, might define a much higher ceiling to be reached through a follow-up curve.

There are more concrete benefits from fitting an S-curve to these data points. The mathematical description of this curve, the logistic function, relates price and performance with time in a quantitative way. Specifying two of these variables dictates a value for the third one. The most obvious place to use such a result is in pricing a new product. Its performance (MIPS) has been measured and is not easy to change. For a chosen introduction date the function determines the MIPS/$ ratio that yields a price value. With such guidelines to decision making, one can aspire to keep doing the same thing—be successful!—and be less subject to errors resulting from today's stressful environment.

But in addition to price setting, the results of such a study provide a means for deciding on corrective actions following delays on product delivery. Product announcements precede shipments by a variable amount of time. Like cash sitting at home, products coming to market late are worth less. How much should a price be dropped following a nonscheduled delay? Alternatively, how much should the performance be improved if we insist upon keeping the same price?

With the logistic-growth function determined, mathematical manipulations (partial time derivatives) can provide answers to the above questions. For the first case we find that the price adjustment following a trimester's delay should decrease exponentially with time. In other words delivery delays hurt the most at the time of introduction. Later on, further delays cost less. This result should not come as a surprise to business people.

What may surprise most people is the second case, improving the performance in order to keep the same price. The performance enhancements required from a trimester's delay go through a bell-shaped evolution. Naturally, as shipments are delayed, the performance must be increased. This increase is larger every year as we approach the middle of the learning curve, which for Digital's low-end computers falls around 1995. From then onward, however, the product will require less and less improvement in its performance for comparable delays. This can be understood in terms of the MIPS/$ growth curve starting to flatten out after 1995, and consequently a slippage in delivery would require a smaller improvement in performance than it would have previously.

The Industry Learning Curve

Most industrial knowledge has been acquired on the job. As a consequence, such knowledge is characterized by unquestionable validity but also by a lack of analytical formulation. Unquestionable truths are intriguing because they signal an underlying law. In my search for guidelines for successful business decisions, I found that mathematical manipulations of the logistic function yield pairs of relationships between price and time, performance and time, or price and performance. While I was carrying out those manipulations I stumbled onto a fundamental piece of industrial knowledge: the economies of scale.

Economies of scale say that the more units you produce of a product, the less it costs to produce a unit. Production costs decrease with volume for a variety of reasons: automation, sharing overhead, reducing material costs through wholesale prices, and general process optimization resulting from experience acquired while producing. A major part of the cost reduction can be attributed to *learning* in some way or another.

The concept of economies of scale is taught qualitatively in business schools by means of the *volume curve,* which shows that costs per unit decrease as a function of the volume of units produced. This curve looks like a decaying exponential, reaching a minimum final value when the costs can be reduced no further. It turns out that this curve is our familiar S-curve in disguise.

The performance/price indicator mentioned previously follows S-curves according to a logistic function. The mathematical expression of this function is a simple fraction that has a constant in the numerator and a constant plus a decreasing exponential in the denominator (see Appendix A). Therefore, its inverse, namely price/performance, is simply a constant plus an exponential. In this form it says that for one unit of performance the price decreases with time down to a final value, beyond which no further decrease is possible. This conclusion based on units of performance can be reproduced by analogy for units of sales.

In other words the volume curve, representing costs per unit, is nothing else than the inverse of the S-shaped learning curve, representing units per dollar. Business persons using volume curves for decades have been dealing with S-curves without realizing it. The fundamental process is *learning*. It can take the form of an S-curve for the variable MIPS/$ or look like a decaying exponential for its inverse, $/MIPS. In either case the law that governs its evolution is natural growth under competition.

GOING WEST

Collective learning by a body larger than a company or an organization is behind such social endeavors as explorations. The great explorers acquired knowledge about remote exotic places and people, a whole new world, and how the earth is round, how to cross the ocean, and how to deal with new types of hardship, diseases, diets, and violence. Later on all of this was recorded in encyclopedias, textbooks, and scientific manuals. Oddly enough, however, the people who carried out the explorations were not men of science or intellectual pursuits. On the contrary, learning was the least of their concerns.

Why then did they feel compelled to do it, and who was behind the learning process?

• • •

It is August 2, 1492, in Palos, Spain. There is feverish activity in the harbor around the boats that will set sail westward tomorrow. Christopher Columbus pauses for a minute on the captain's deck of the Santa Maria. He leans forward over the railing, gazing down at the deck below, watching his crew at work.

He is an old master of the seas. He has no apprehension; he is eager to set sail. The other side of the ocean draws him like a siren. But he realizes that he is not the first one to be possessed like this. Many captains have journeyed westward before him, but none of them came back. How many? he wonders. Over the centuries, how many ships sailed off and never sent back a single word, not even a signal?

He is burning with desire to succeed. What are his chances? If those who perished before him number half a dozen to a dozen, maybe it was their fault. It is not impossible that ten captains in a row made bad mistakes, showed poor judgment, or just had bad luck. But if there were hundreds before him? If hundreds or thousands of captains before him failed trying to sail west, why should he succeed?

He raises his glance to the dark horizon over the ocean. It couldn't have been so many. If hundreds of boats like his had been lost, it would have become a legend. Kings would have gone bankrupt! Or it must have been a long, long time ago, so long ago that boats would have been primitive, flimsy, inefficient. With boats of inferior design, the explorers could not have survived.

He squares his shoulders, takes a deep breath, looks up at the sky, and reenters his cabin. There is work to be done. Tomorrow he is sailing.

·   ·   ·

Was Columbus's reasoning right? Were there only a few attempted expeditions westward before him? Given that there is no accurate documentation, is there any way one can estimate how many expeditions westward failed before Columbus's voyage?

Let us assume that explorations represent some form of learning and therefore conform to the S-curve configuration that is typical of the learning process. Since learning implies that there is someone who learns—the child in the case of vocabulary, the scientific community in the discovery of the chemical elements—the voyage of Columbus together with all other explorations constitute the learning process in which Europe discovered the new world. From the successful Western Hemisphere explorations that have been well documented, we can draw Figure 2.2 below, showing the cumulative number of explorations in ten-year intervals, following Columbus's voyage.[3]

The data themselves suggest that we may be tracing the rapid ascent and the leveling off of the middle and late part of an S-curve. Its origins are back somewhere in the fourteenth century. Accordingly, the fit to this kind of data is made by allowing an early missing part of unknown size. The best S-curve that can be made to go through the data is the line shown in the illustration.

Since S-curves go to zero asymptotically (that is, they become zero only when time equals plus or minus infinity), we take as a nominal beginning the time when 1 percent of the ceiling is reached. In any case, below this level we are talking about a fraction of an exploration. A "beginning" defined in this way points to the year 1340. This date can be considered as the time when the exploring process originated, the moment when Europe's need to explore the West was born. For the best agreement between curve and data, the early-missing-data parameter has to take a value around fifteen.

The conclusion is that there must have been about fifteen attempts

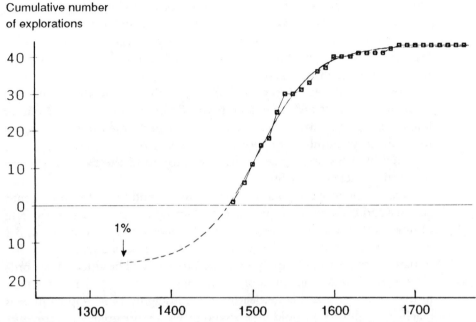

WESTERN HEMISPHERE EXPLORATIONS

Cumulative number
of explorations

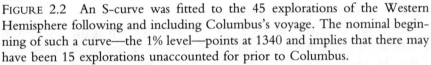

FIGURE 2.2   An S-curve was fitted to the 45 explorations of the Western Hemisphere following and including Columbus's voyage. The nominal beginning of such a curve—the 1% level—points at 1340 and implies that there may have been 15 explorations unaccounted for prior to Columbus.

before Columbus that failed while trying to explore the West, the first one dating back to around 1340. This kind of *back*casting—if somewhat daring—is not different from the usual *fore*casting approach. We are dealing with a pattern similar to that of the growing vocabulary of an infant if we only look at the data pattern upside down and let time run backward. The uncertainties on the number fifteen can be estimated (see Appendix B) to be plus or minus three.

MESOPOTAMIA, THE MOON, AND THE MATTERHORN

Learning, discovering, and exploring are activities of similar nature and are often encountered together, as in the field of archaeology. Archaeologists dig and analyze in order to learn about our past, at times in remote places on Earth. Mesopotamia is such a distant region, rich in buried secrets of remarkable ancient civilizations. This itself is no secret; scholars have been excavating the region for centuries. The rate of excavations can be taken as a measure of the archaeologists' interest, reflecting the amount of knowledge obtained. Treating the process as learning, it should be possible to forecast the number of excavations remaining to be undertaken.

The way to do this is to fit a natural-growth curve to the cumulative number of excavations. The ceiling will define the size of the niche of knowledge that is slowly being exhausted. I performed this exercise on the cumulative number of major excavations carried out in that region as reported by a historical atlas.[4] The curve followed the data closely and practically reached the ceiling.

Excavations in Mesopotamia started in the middle of the nineteenth century and grew to reach peak activity during the 1930s. Later, the rate of new excavations in the region slowed. It is a searching process similar to the one of finding needles in a haystack, and the number of needles remaining to be found slowly dwindled to a few. The fitted curve predicts there are practically no more major new endeavors to be undertaken in that region. By and large, learning from digging in Mesopotamia has reached a ceiling. Like goldminers who find no more gold, archaeologists are turning elsewhere for their searches.

The natural-growth curve can also be applied to chart the progress of explorations of the moon. Moon explorations came and went in less than a decade, leaving behind an S-curve of visits to our nearest neighbor. That is what we get if we plot the cumulative number of American moon missions. The total number of launches to the moon (both manned and unmanned) is only fourteen, which makes the statistical fluctuations rather visible. Therefore, one can expect some scattering of the data points around the curve. For example, the second expedition came a little too early, while the fourth one was a little late. But on the average the data points follow the S-curve pattern closely all the way to the ceiling (Appendix C, Figure 2.1). The moon knowledge "niche" was exhausted by 1972, after which no expeditions were undertaken. America's preoccupation with the Earth's natural satellite began, intensified, and subsided in a total of eight years, like a grandiose fad.

To a certain extent expeditions to conquer high mountains follow the same pattern. They start slowly but soon reach a feverish "gold-rush" period. But mountain climbing, unlike gold-panning, displays an interesting aftermath. It is something I first noticed with the expeditions to the Matterhorn, the second highest mountain in Europe after Mont Blanc

(4400 meters compared to 4800 meters). My observation highlights one of the eventualities to be expected after the completion of an exploration process.

.   .   .

Once upon a time in an alpine chalet with skiing conditions officially declared not practicable, I dug into a pile of old books stacked away in a corner collecting dust and absorbing the smell of wood smoke. Among them I found a calendar of expeditions on the Matterhorn, the Swiss alpine shrine that rises like a huge pyramid, presenting three different faces to mountaineers. It has been explored methodically from all sides. When the easiest side was successfully climbed, mountaineers turned to the more unfriendly ones, in order of increasing difficulty. Each new, more difficult ascent required perseverance before it was achieved and claimed its share of accidents. Each climb contributed another data point to an S-curve that started unfolding before my eyes. The natural-growth pattern of the learning process broke down, however, before the curve was completed. As time went on more and more people climbed the same old paths. Today, hordes of tourists climb the mountain every day. In contrast to the smooth bends of the early curve, which described the exploration and learning process, today's pattern of the cumulative number of expeditions depicts a steep straight-line trend reflecting up to a hundred climbers per day in well-organized groups.

I struggled for a while trying to understand the meaning of the graph in front of me. The answer came to me with an understanding of the difference between exploration and mere travel.

.   .   .

Tourists are not explorers. Tourists go where many others have gone before. For an undertaking to be qualified as an exploration, it must involve learning through doing something *for the first time.* The more times it has been done before, the more reason for this activity to be classified as tourism. Both exploration and tourism involve learning, but of different quality and scale. Exploration consists of many events that conform to the pattern of a learning process—for example, the S-curve that describes Europe's learning about the Western Hemisphere. Tourism is made of isolated events with no connection to one another. The only ones who learn from them are the individual tourists.

We will come back to look at the relationship between S-curves and tourism in more depth in Chapter Ten, but what we can conclude at this time is that exploration, like the learning process itself, follows the S-curve pattern of natural growth. Therefore, it becomes possible to estimate the time of its origin, if it is not documented, as well as the time that it will slow down and stop, even if that time is well in the future.

3

Inanimate Production Like Animate Reproduction

What do computers and rabbits have in common? They can multiply. What's more, they do it in just about the same way!

. . .

The Belgian mathematician P. F. Verhulst must be considered the grandfather of S-curves. He was the first to postulate a law containing a limitation for the growth of populations, back in 1845. He argued that a given niche has a finite capacity for maintaining a population and that the growth rate should drop when the population size approaches the niche capacity.[1] Later on many people developed this subject further. Two men of science built a mathematical formulation linking the population of a species to the one on which it feeds. The system of equations that bears their names became widely accepted by biologists and ecologists as the foundation of competitive growth models. These two men were Alfred J. Lotka and Vito Volterra.

The careers of both Lotka and Volterra culminated during the early decades of our century. At that time the borderlines between disciplines were not felt as strongly as today. Lotka, for example, a professor of physics at Johns Hopkins University, would not hesitate to involve statistics, biology, mechanics, and the understanding of consciousness all in the same work. Volterra, an Italian mathematician and professor at the universities of Pisa and Rome, contributed to the advancement of mathematics, physics, biology, politics, and Italian aviation during World War I. In the 1920s Volterra and Lotka cast the predator-prey relationship into a set of equations.[2] The work, now known as the Volterra-Lotka system of differential equations, draws on Verhulst's original description of natural growth for a species population.

There have been numerous observations that species populations do indeed grow along S-curves. It was reported early in this century that under controlled experimental conditions populations of fruit flies grew along S-shaped curves.[3] The fly population followed the smooth theoretical curve quite closely all the way to the end (Appendix C, Figure 3.1). The intriguing question, which Lotka was the first to ask, is whether this theory can also describe the growth of *inanimate* populations.

There are, in fact, similarities between the growth of rabbits in a grass field and automobiles in society. In both cases growth is capped by a limited resource that becomes scarce as the population competes for it. For rabbits it is grass, for cars it may be space, which is becoming increasingly precious in city centers and on highways.

During the early phases of a rabbit population, the growth is rapid—exponential in mathematical terms. If each couple has two offspring, the population doubles every mating cycle. The element of rapid, exponential-like growth during the early stages is also observable in car populations. In an affluent society one car brings another. In a household in which both parents have a car, the daughter asks for hers as soon as the son gets his. The number of cars owned by the neighbors may play a role, and it is not rare to find individuals owning several types of cars to match a variety of personal uses. The more we become accustomed to cars, the more we need them—at least it seemed that way some time ago.

The conceptual similarity between the way rabbits colonize a grass field and automobiles "colonize" society points to a common underlying law. The difference lies with the time scales. It may take rabbits months or years to fill up their ecological niche, while it takes cars decades to fill up their niche in society. Confirmation of this hypothesis, however, can only come from the real data.

The data on cars registered officially in Italy since 1955 show that the annual number grows along an S-curve identical to that of fly populations mentioned earlier (Appendix C, Figure 3.1). There are two differences. One is the time scale; instead of reaching the ceiling in a matter of weeks, it takes decades. The other is that having a much larger sample of cars than flies, the statistical fluctuations are smaller and the agreement between the data points and the curve is even more striking for cars than for flies.

It is remarkable that world-shaking events like the two oil shocks—one in 1974 and the other in 1979—left no mark on the smooth pattern of the growing number of cars, vehicles so intimately linked to the availability of oil and the general state of the economy. Many indicators, including oil prices, flared up and subsided during this time period. Car production echoed the economic shocks, but not the number of cars in use. During difficult times people deferred buying new cars to save money in the short term. They simply held on to their old cars longer. The number of cars in use kept growing in the normal pattern. This number is dictated by a fundamental social need and is not affected by the economic and political climate. In this light, events that produce big headlines may shake the world, but they do not significantly influence the pattern of growth of the number of cars on the street.

I became involved in fitting S-curves to populations of man-made products while I was looking into the possibility of computer sales following such growth. The first computer model I tried turned out to be a showcase, one of Digital's early successful minicomputers, the VAX 11/750. The cumulative number of units sold is shown at the top of Figure 3.1. An S-curve passes quite closely to all twenty-eight trimesterly data points. In the lower graph we see the product's life cycle, the number of units sold each trimester. The bell-shaped curve is the life cycle as deduced from the smooth curve at the top.

When I produced this graph in 1985, I concluded that the product was *phasing out,* something that most marketers opposed vehemently at the time. They told me of plans to advertise and repackage the product in order to boost sales. They also spoke of seasonal effects, which could explain some of the recent low sales.

The data points during the following three years turned out to be in agreement with my projections. To me this came as evidence that promotion, price, and competition were conditions present *throughout* a product's life cycle and have no singular effect. The new programs that marketers put in place were not significantly different from those of the past and therefore did not produce a modification of the predicted trajectory.

Still today I find individuals claiming that the phasing out of the VAX 11/750 was an unfortunate event precipitated prematurely by the launching of MicroVAX II, a computer as powerful but at one-third the price.

A SUCCESSFUL COMPUTER PRODUCT MAY FILL ITS NICHE LIKE A SPECIES

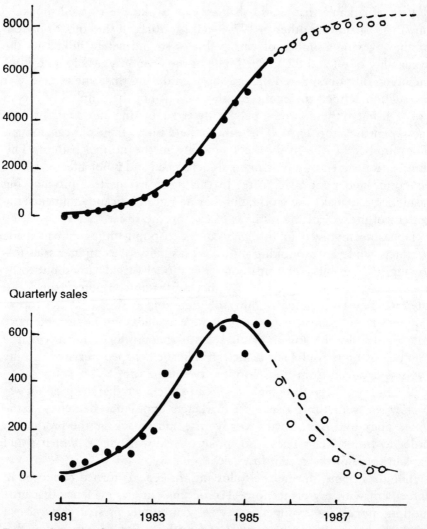

FIGURE 3.1 At the top, growth of cumulative sales in Europe for Digital's minicomputer VAX 11/750. The S-curve shown is a fit carried out in 1985. The dotted line was a forecast at that time; the little circles show the subsequent sales. The product's life cycle at the bottom has been calculated from the figure at the top.

Their arguments are irrational. Like early twentieth-century physicists resisting quantum mechanical concepts, they are having a hard time with the facts of Figure 3.1, which spelled out the fate of the VAX 11/750 three months before MicroVAX II was announced.

Since that time I have tried to fit S-curves to many products. Often I have been disappointed, particularly recently when computer models appear in rapid succession with little differentiation. The 11/750 was a long-lived, well-positioned product that had its own market niche. Today's models overlap and often share market niches with other products. Life cycles are too short and behave too irregularly to be fitted to the theoretical curve. What is emerging as a better candidate for a growth curve description is a whole family of models, a generation of technology.

But this was also considered by Lotka. He linked the growth of a population to the growth of an individual. A colony of unicellular organisms, regarded as a whole, is analogous to the body of a multicellular organism. Imagine a cup of soup left on the table for several days. Bacteria will grow in it subject to food limitations. The transformation changes a "grain" of the substances in the soup into a "body" for the bacteria. H. G. Thornton has studied the case and has shown that bacterial growth (measured by the surface area occupied) follows the same familiar growth curve pattern (Appendix C, Figure 3.2).

Whether it is bacteria in a bowl of soup, rabbits in a fenced-off grass field, or automobiles in society, there is a progressive transformation of a limited resource into a resident population. If the multiplication is exponential—situations in which one unit brings forth another—the overall pattern of growth is *symmetric*. It is this symmetry that endows S-curves with predictive power; when you are halfway through the process, you know how the other half will be. The symmetry of the logistic function can be mathematically demonstrated, but it can also become evident intuitively.

· · ·

There is road construction going on next to a thirteenth-century church in Paris. In front of a magnificent stained-glass window, red steel barrels filled with small stones line the work site. A teenage vandal passing by at night looks at the glass mosaic, which contains over a hundred small stained-glass pieces. The bright street light reveals a black hole; one piece is missing. The contents of the barrel give him an idea, and he throws a stone at the hole. He misses, but his projectile knocks out another small pane; now there are two holes. He liked the sound it made, and so he throws two stones simultaneously. That brings the number of holes in the window to four. He decides to develop this game further, each time throwing a number of stones equal to the number of holes and giving up any pretense of aim. The random hits make the number of holes increase exponentially for a while (4, 8, 16, and so forth), but soon some stones fall through holes causing no further damage. The vandal adjusts the number of projectiles per throw according to the rules of his game. But as the number of the glass pieces still intact diminishes, he realizes that his rate of "success" is proportional to this number. Despite throwing tens of projectiles at once, his hit rate is reduced by a factor of two each time.

Inside the church, under cover of darkness, a tourist is taking advantage of his sophisticated video-recording equipment in order to bring home pictures that tourists are not normally allowed to take. Among other things he films the complete demolition of the stained-glass window. Aware that he is trespassing, he does not interfere.

Back home he shows his friends the beautiful vitraux before and after. Faced with the dismay and indignation of the audience, he plays his tape backwards, and the masterpiece is reconstructed piece-meal. The audience watches the empty honeycomblike lead frame-work become filled with sparkling colored glass, first one piece, then 2, 4, 8, 16, and so forth, until the exponential growth slows down. Someone points out that the growth pattern in the recon-struction is identical to the sequence of hole appearances in the original scene of vandalism.

•   •   •

This story provides an intuitive template for population growth in a space of limited resources. Rabbits, like automobiles, grow exponentially in the beginning. One brings forth others until competition for a diminishing resource makes the difficulty of obtaining it inversely proportional to the amount of the resource left. The simplest natural-growth pattern is a symmetric one.

The same pattern can clarify why and when chain-letter activities break down. A chain letter demands that each recipient send out a number of letters, usually more than two. If this number is ten, the population of letters grows exponentially, but faster than the number of projectiles in the above story, because in each step it is multiplied by ten instead of two. We will successively see 1, 10, 100, 1,000, 10,000, and so forth, letters being mailed. The chain will start breaking down as soon as the number of letters becomes significant when compared to the number of potential participants.

The bacteria in the bowl of soup can be viewed both as a population and as an organism. The size of living organisms, from trees to humans, is related to the size of the cell populations that make them up. Cells multiply simply by dividing into two. Thus, the size of organisms grows exponentially in the beginning, but their growth is subject to a limitation, the genetically predetermined final height. Can it be predicted?

DO NOT TRY TO FORECAST WHAT YOU ALREADY KNOW

Genetic predetermination makes prediction possible. A black couple can forecast the skin color of their baby with high confidence even if, in such situations, we prefer words less loaded than "forecast," "confidence," and "genetic predetermination."

· · ·

"What do you do at work, Dad?" my daughter asked me one day.

Having been working on forecasting, I wondered if at the age of ten she knew what the word meant. "I tell the future," I answered.

"You can do that?" she wondered in disbelief.

"For things that are growing, I can tell how big they will eventually be," I replied.

"Well, I'm growing," she said. "Can you tell how tall I will be?"

"I could try," I said, "if I knew how much you have grown every year so far. I only have a few measurements from when you were a little baby."

"At school they measure us every year from kindergarten," she said. "The nurse keeps a record of all these numbers on a card."

The next day she came home with a photocopy of her school card containing height and weight measurements for the last six years. "Here you are," she said. "Now how tall will I be?"

I turned on my personal computer and typed in the numbers. She stood next to me radiant with doubting expectation.

The data set was not complete. There was a gap between ages one and four. I tried to compensate by adjusting the weights accordingly.

At the end, I said rather triumphantly, "You will be five feet six inches."

First she had to appreciate how tall that was. We took the tape measure, made a mark on the door, and looked at it from different angles. She seemed pleased. But soon she was concerned again. "How sure are you?"

She was only ten; she knew nothing about measurement errors, statistics, standard deviations, or uncertainties from fitting procedures. But her instinct was warning her about the magnitude of the uncertainties in determining a final plateau from early measurements. I didn't want to disappoint her, but I didn't want to lie to her, either. I made a quick calculation "sprinkled" with optimism, using Table III from Appendix B.

"You will be five feet six to within ten percent," I decided.

"What does ten percent mean?" she asked.

"It means you will most probably be between four feet eleven and six feet one," I offered timidly.

A moment of reflection, some visualizations, and then she turned away, disappointed and mumbling, "Big deal."

· · ·

In fact, in addition to my daughter's measurements, the card of the school nurse had curves already traced out through an accumulation of statistical data for girls of this region over many years. These curves looked like the upper half of S-curves. My daughter's data points fell systematically on the curve, the final height of which was five feet and six inches at around sixteen years of age.

Fitting growth curves to measurements of children's height may not be the recommended approach for determining their final size. More accurate methods exist that are based on the fact that children have been amply observed and measured while growing, and by now there is a wealth of tables and curves that point to the final height with satisfactory accuracy.

There may be organisms, however, that grow *for the first time* in history so that no record can be consulted on how it happened before. In such cases, forecasts are useful in spite of significant uncertainties. Let us consider the great American railway network as an example of a one-of-a-kind "organism." Should its growth follow the familiar pattern of a population? Is its final size predictable?

The answer can be found in Alfred J. Lotka's *Elements of Physical Biology,* published in 1925, at a time when railways were still the primary means of transportation.[4] Lotka presents a graph of the total mileage of railway tracks in the United States from the beginning of railway history until 1918, when mileage had reached 280,000. Despite not having computers at that time, he superimposes on the data a theoretical S-curve that passes very closely to all points and shows a 93 percent saturation of the U.S. railway "niche" by 1918, and a ceiling for a total mileage of about 300,000. I looked up the amount of railway track laid since 1918; it is about 10,000 miles, which largely confirms Lotka's prediction. Confirmation of his prediction is all the more impressive considering that seventy years ago rail transport was widely accepted as a young industry with a promising future.

To highlight the analogy, Lotka includes in his work a graph of the growth of a sunflower plant, a typical case of a multicellular organism growing along a population curve (Appendix C, Figure 3.3). The two processes seem identical except for the scales. The sunflower seedlings achieve their maximum size (eight feet six inches) in eighty-four days, while the U.S. railway track mileage following the exact same curve reached maturity (290,000 miles) in 150 years. Something as inanimate as man-made ironworks has grown in a way typical of living organisms. In that context it seems particularly appropriate that rail terminology makes use of terms such as tracks, trunk lines, branches, and arteries.

MAN-MADE DINOSAURS

The railway infrastructure is not unique in having followed a natural-growth pattern. Throughout history and over a broad range of time scales, people have built structures that grew in numbers like species, with well-defined life cycles.

• • •

King Kong is breathing heavily, lying unconscious on the steel floor of the enclosure that is confining him, stunned by the rocket-injected tranquilizer which allowed humans to catch him and transport him to civilization. Transporting King Kong is something easier said than done, however. According to tradition, the giant ape is five stories tall and weighs hundreds of tons. What means of transportation could do the job?

A supertanker, a man–made dinosaur.

• • •

In the original movie King Kong was transported in the hold of a conventional cargo vessel, which would have been impossible. Conveniently, when the movie was remade there were supertankers available, which could easily accommodate the huge animal. If the movie is made again in a few decades, these vessels will no longer exist, and King Kong may very well be forced to swim to civilization.

Supertankers are sea vessels with a carrying capacity of more than three hundred thousand tons. A global fleet of 110 supertankers was constructed in the 1970s; construction ended less than ten years later. The wave of supertankers came and went like a fad. Its lifetime was so short that one may be tempted to consider the whole affair an experiment that failed. Smaller tonnage ships seemed to have better survival characteristics. Thus constrained by a competitive squeeze, the supertanker population followed the pattern of a species' growth as beautifully evidenced by a precisely outlined S-curve (Appendix C, Figure 3.4). Once the ceiling was reached, construction ended, and many such ships have already been decommissioned. Monster supertankers are heading for extinction, without being replaced.

Over the centuries society's spotlights have shone on structures of even more extravagant scale. Two such examples are Gothic cathedrals and particle accelerators. Like supertankers, their life cycles are also characterized by a natural-growth curve. As the populations of both reach the ceiling, obsolescence sets in, and society no longer invests funds or energy in them.

Construction of Gothic cathedrals began at the end of the eleventh century. These endeavors represented formidable undertakings, and their realization became possible only through the cooperation of groups with disparate identities: clergy, architects, masons, royalty, and peasant communities. These groups, with mostly conflicting interests, collaborated for decades to complete the work. The interest in such monumental buildings grew and spread over Europe spontaneously and rapidly. During the twelfth century cathedral cornerstones were being laid in cities across Europe in quick succession. Many cathedrals were being built simultaneously. It was not the magnificence of a completed work that inspired new undertakings. New ones were begun before the old ones were completed. Europe seemed to have become a fertile ground for a population of cathedrals. During a period of 150 years the rate of appearance of new cathedrals peaked and then declined. By 1400 no more cathedrals were being started; the population had reached its ceiling like a species that has filled up its ecological niche.

In gathering the data concerning the construction of cathedrals, I had to distinguish Gothic cathedrals from large churches. I also had to estimate a cornerstone date for old churches that were "upgraded" to cathedrals later. Such uncertainties may be responsible for some scattering in the data points above and below the fitted natural-growth curve. In any case one can safely draw the overall conclusion that Europe's need for Gothic cathedrals was largely satisfied in a "natural" way by A.D. 1400 (Appendix C, Figure 3.5).

In more recent times physicists have built laboratories of extravagant dimensions called particle accelerators or atom smashers. Some years ago in an inaugural speech, the director general of the European Center for Nuclear Research (CERN) in Geneva, where the world's largest accelerator is housed today, referred to particle accelerators as the cathedrals of today. This image suggested among other things that an end to this "species" should also be expected, so I decided to study their evolution and probe into their future. Atom smashers proliferated after World War II to produce a population whose growth followed the same pattern as supertankers and cathedrals did before. The historical data I collected in this case consist of the number of particle accelerators coming into operation worldwide. The S-curve fitted on them approaches a ceiling in 1990 (Appendix C, Figure 3.6).

Each accelerator counts as one even if the size of the device varied significantly over the years. The early setups were of modest dimensions. The Cosmotron of Brookhaven National Laboratory, where preliminary work for the discovery of the antiproton was carried out in the 1950s, was housed inside one big hall. At CERN, the ring of the electron accelerator LEP was completed in 1989. It measures eighteen miles in circumference, lies deep underground, and passes under suburbs of Geneva, several villages, and the nearby mountains. The accelerator currently under construction in Texas, the Superconducting Super Collider (SSC), will also occupy an underground tunnel, but with a circumference of almost fifty-four miles.

One may think that accelerators like the Cosmotron and LEP, being so different in size, should not both count as one in defining the population. On the other hand, building the Cosmotron with the know-how of the early 1950s was no lesser a feat than the ring of LEP in the late 1980s. Particle physicists gave the best of themselves and were rewarded comparably in each case. T. W. L. Sanford studied the succession between accelerators and found that as soon as a new, more powerful accelerator came into existence, physicists, experiments, and publications shifted from the old to the new. This changeover was not abrupt, however, but along smooth S-curves.[5]

The reason the size of accelerators has grown to practically absurd dimensions is that we have progressively exhausted the utility of the fundamental principle of electron accelerators: the radio-frequency electromagnetic cavity used for acceleration. Every new horizon of particle physics research requires higher particle energies, more acceleration, and more radio-frequency cavities. In the end it is purely the size, with the associated expenses, that will render the process obsolete.

The theoretical growth curve fitted to the data points indicates that there may be one or two more accelerators still to be built worldwide. But by and large the era of high-energy physics particle research as we have known it is coming to a close. It is also of interest to note that the nominal beginning of the curve (the 1 percent of maximum level) does not point at Ernest Rutherford's famous experiment in 1919 when he first used particles to explore the structure of the atom. The beginning of the accelerator era is pinpointed at the outbreak of World War II, but "technicalities" during the war years delayed the actual construction of laboratory sites for about ten years. The data do in fact show an early catching-up effect. Seen in retrospect, Rutherford's seed was given a chance to sprout thanks to the intense activity on nuclear research during the war years. One may even want to see the subsequent demonstration of overwhelming power in nuclear energy as the potent fertilizer that made accelerators grow to gigantic dimensions.

In all three examples—the construction of supertankers, cathedrals, and particle accelerators—competitive growth was the law in effect. All three were relatively short-lived endeavors. Extinction was probably precipitated by competition-related hardships attributed to their exorbitant size. Size, however, is not necessarily a reason for early extinction. The real dinosaurs lived longer than most species on Earth, and they did not become extinct by slowly phasing out due to their size. The most popular theory today is that dinosaurs died unnaturally due to a violent change in the environment, probably caused by a collision between the Earth and a large meteorite, which raised enough dust in the atmosphere to eclipse the sun for a number of years, thus killing most vegetation. Around 70 percent of all species disappeared at that time, including herbivorous dinosaurs. The difference between real and man-made dinosaurs is that the former grew slowly, adapting continuously with a life history greater than 100 million years and consequently well suited for many millions more. The man-made ones grew and declined rapidly. Life cycles are generally symmetric, as exemplified by proverbial wisdom: early ripe, early rot.

Mama Earth

The building of impressive structures requires raw materials, energy, and wealth. In addition to living off the land, humans are extracting from it substances of exceptional properties sometimes in incredible quantities. Gold is such a substance, oil is another. Originally of entirely different uses, these two have often served interchangeably. In spite of a large difference in the time of appearance, they have both at some time or another served the purposes of war, blackmail, prosperity, and power.

For both cases extraction is a one-way process. Oil, once burned, does not return underground to replenish itself. Similarly for gold; once processed and owned, it rarely finds its way back to earth. Furthermore, natural replenishment takes much too long and cannot make up for the usage. For that reason speculation about and estimation of reserves have become a popular game, and there are all too frequent revisions and updates.

Although they are inanimate, we can look at the surfacing of these substances as if they were "populations" growing to fill—or empty—a "niche." The niche may be the amount of the precious material Mother Earth has in store underground for us. Alternatively, the niche may simply be the amount of the substance in question for which we have a well-defined need. The argument that human desire for energy and gold is insatiable may be simplistic. Oil as a primary energy source is already losing ground to natural gas and nuclear energy (see Chapter Seven), and that is happening well before reserves run out. As for gold, there have been micro-niches of different usages coming in and out of fashion all along: art, dentistry, electronics, the space industry, and others. Materials that can substitute for gold—sometimes better suited for a particular use—are continuously invented. It is not clear that the need for gold will continue to increase indefinitely.

If the production of gold and oil proceed so as to strike a continuous balance between supply and demand, it must follow a *natural* path. Evidence that a path is natural is its similarity to an S-curve, which would indicate that the rate of growth is proportional to the part of a niche still remaining unfilled. Production of gold and oil is indeed found to behave like this.

I studied the world production of gold by plotting the cumulative amount of the metal extracted since 1850.[6] This date can be thought of as the beginning of contemporary gold mining, since the prior annual rate of world production was more than a factor of ten smaller. However, a different scale natural-growth curve would probably be suitable to describe gold production from antiquity to 1850.

I found the data points on the graph to accurately outline the first 45 percent of an S-curve. The projected ceiling of gold production is approached toward the end of the twenty-second century. Even if gold is no longer being found in America, the world production will keep increasing on the average to reach the maximum rate of annual extraction around 2025, which will thus become a kind of *golden era*.

The case of oil is somewhat different. Oil started being produced commercially in 1859, but production picked up significantly only in the early twentieth century. From the beginning, extraction of oil stimulated exploration for new reserves. Exploration being expensive, however, it was pursued only to the extent deemed necessary at the time. Figure 3.2 shows both production and discovery of reserves for the United States. The historical data (depicted by thick lines) represent cumulative production and cumulative discovery of reserves.[7] Oil production is a smoother process than discovery, which features bigger fluctuations due to the inherent randomness associated with a search. Both sets of data permit excellent fits to S-curves (depicted by thin lines).

The two curves turned out to be remarkably parallel, with a constant separation of ten years. Such a rigid linkage between production and discovery, witnessed for ninety years, is proof of an underlying regulatory mechanism based on a closed feedback loop.* In feedback loops causality works both ways. Finding more oil may result in increased production, or alternatively, increases in production may provoke intensification of efforts to find more oil. In any case the strict regulation says that we discover oil reserves ten years before we consume them, not earlier or later. Naturally, the effort going into searching for oil may increase as we exhaust easily available supplies; one alternative would be to search more deeply. Finally, production may slow down as a consequence of stiffening conditions for finding new reserves. Whatever happens, the regulation observed in Figure 3.2 guarantees ten years of reserves at any given time.

---

* A feedback loop is a cyclical process in which corrective action is taken in response to information about an earlier state. A typical feedback loop is a heating system controlled by a thermostat.

OIL DISCOVERY AND PRODUCTION GO HAND IN HAND

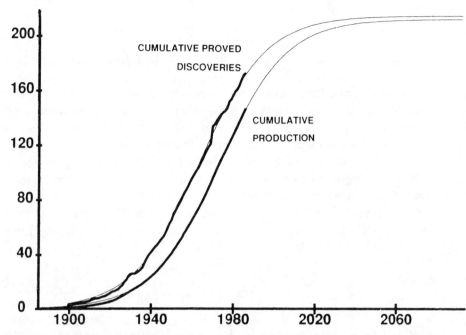

FIGURE 3.2   Data (thick lines) and S-curve fits (thin lines) for oil discovery and production in the United States.

It should be emphasized that no conscious decision has been involved in producing this equilibrium. On the contrary, so-called experts, misusing statistical information, have often forecast imminent doom, with oil shortages and even utter depletion within the next few years. There have been no advocates for the regulatory mechanism depicted in Figure 3.2: The more you milk the reserves, the more reserves will be made available to you.

Even if oil discoveries are finally slowing down in the United States, the rest of the world has been poorly explored so far, both in area and in depth. In fact, estimates by experts on the planet's unfound reserves keep increasing with time. In 1977 they ranged from a factor of 3 to a factor of 10 of the then known reserves.[8] No matter what this factor turns out to be, chances are we will stop using oil *before* we run out of it on a global level. It has happened more than once before. We moved from wood to coal well before running out of wood. Similarly, the decline in the use of coal in favor of oil was not driven by scarcity. As we will see in more detail in Chapter Seven, the driving force in substitutions between primary energies is not scarcity but the different economies of scale involved in the technologies of each energy source.

DID CHRISTIANITY BEGIN *BEFORE* CHRIST?

Production along population curves includes the widest range of human endeavors. The cases of oil and gold presented above have been relatively recent activities. A process that reaches much further back into the past is the canonization of Christian saints. The church offers rich and fairly accurate records of these events, accumulated and maintained by people who were in general educated and who adhered to a continuous tradition lasting almost two thousand years.

The influence of the church and the importance of religion have not been constant over the years. It was Cesare Marchetti's supposition that the rate of canonization might be used as an indicator of society's involvement with religion throughout the centuries. He claims to have found two broad peaks in the rate of canonization across the history of Christianity. When I myself located a reliable and complete list of saints, I wanted to check his result.[9] I tabulated the number of saints per century and indeed observed two peaks. In Figure 3.3 we see centennial points representing the cumulative number of saints as a function of time. I was able to fit two S-curves that together seem to describe well most of the historical data points.

The two waves have comparable lengths, around one thousand years each. Such a double structure may reflect the well-established notions of the patristic and Thomistic ecclesiastical waves, which correspond roughly with these chronological periods. The former represents the influence of the early fathers of Christianity, while the latter is attributed to Thomas Aquinas, who was born in 1224 and died in 1274.

The two curves do not cascade in a continuous way. The first wave ends around the eleventh century, but the next wave does not start until the thirteenth century. During the time interval between them, canonization proceeds at a flat rate—constant number of saints per century—as if this period belongs to neither group. It is easy to offer explanations a posteriori, and therefore they should not carry much weight, but it is interesting to notice that this time of ill-defined sainthood identity coincides with a rather unfitting—brutal and militaristic—religious expression, the Crusades. The ceiling of the second wave is to be reached

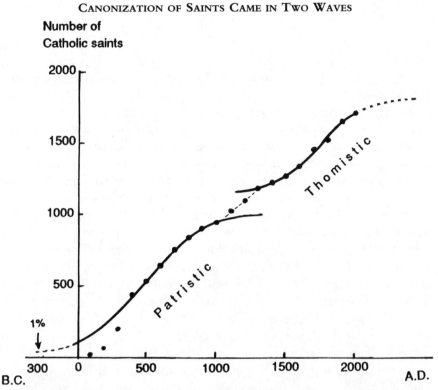

Canonization of Saints Came in Two Waves

FIGURE 3.3 The two S-curves have been fitted separately to the historical windows 400–1000 and 1300–1970. The first curve, called patristic, is back-casted to a nominal beginning around 300 B.C. The second curve, called Thomistic from Thomas Aquinas (1224–1274), is forecasted to reach a ceiling soon. Between the tenth and the thirteenth centuries, canonization proceeds at a flat rate, indicating a period that does not belong to a natural-growth process.

toward the end of the twenty-first century. One wonders if some form of eccentric religious behavior is to be expected then, too.

The most interesting observation in Figure 3.3 concerns the very beginning of the first curve. The fit deviates from the early data points. The rate of growth in the number of saints during the first few centuries A.D. seems much faster than the more natural rate observed later on. One may interpret this as a "catching-up" effect. People had been expecting a savior for a long time. When one finally came, a certain amount of pent-up energy was released, resulting in an accelerated rate of sainthood attribution.

Had canonization proceeded according to the more natural rate observed from the fourth century onward, it would have started *before* the birth of Christ. If the patristic curve is backcasted, the time when its level equals 1 percent of the ceiling can be taken as an approximate beginning of the process of canonization of Christian saints. From Figure 3.3 we see that this beginning is around the third century B.C., implying that Christianity starts *before* Christ!

Here is Marchetti's explanation.[10] Many speculations have been made about Jesus' whereabouts between the age of fifteen and thirty. The Judaic schism of Essenes was active at the time in that region and, as individuals or in brotherhoods, demonstrated asceticism and extraordinary piety from sometime during the third century B.C. until the first century A.D. Information from the Dead Sea scrolls indicates that their doctrine contained many of the essential elements found later in the Christian message. It is possible that Jesus matured among them.

The backcasting of the patristic curve can be justified if we make two assumptions. First, that there is *one* general S-shape for cultural waves so that the two curves in Figure 3.3 must be similar. And second, that saints and martyrs belonging to the same general line of faith (such as Essenes and Christians) should be grouped together. If these assumptions are made, the part of the curve backcasted before the appearance of Christ provides numbers and dates for individuals who should have become saints *instead* of some hastily canonized early Christian ones. Such deductions might serve as guidelines for a targeted historical research.

Such accelerated growth is often observed at the beginning of an S-curve. The starting up of natural growth might be impeded for a variety of "technical" reasons. But once the growth process is under way, it proceeds faster than normal for a while to make up for the time lost. We will see many more examples and explanations of this phenomenon later on in Chapter Ten.

The growth of a population—be it rabbits, cars, computers, bacteria, human cells, supertankers, Gothic cathedrals, atom smashers, barrels of oil, or Christian saints—may display the S-shaped pattern characteristic of natural growth. Conversely, if a population evolves along an S-curve, the growth process it follows must be a natural one. In either case the life cycle of the process will not remain incomplete under normal conditions. Consequently, missing data can be guessed both in the future (forecasting) and in the past (backcasting) by constructing the complete symmetric S-curve.

4

The Rise and Fall of Creativity

The decade of the sixties witnessed cultural waves that rocked American society with considerable frequency and intensity: rock 'n' roll music, protests against the war in Vietnam, the youth rebellion, and the civil rights movement. Less sharp but equally permeating were more intellectual influences such as Marshall McLuhan's "the medium is the message," Buckminster Fuller's cosmology, and a galloping epidemic of esoteric and spiritual teachings, including the notion that all physical ailments have psychosomatic causes. A quarter of a century later the passage of time has diminished the amplitude of exaggerated reactions to these cultural upheavals. Today, psychosomatic treatment for a torn tendon may appeal to only a few, but a correlation between high productivity and a good state of mental and physical health would not surprise most people.

Cesare Marchetti was the first to associate the evolution of a person's creativity and productivity with natural growth. He assumed that a work of art or science is the final expression of a "pulse of action" that originates somewhere in the depths of the brain and works its way through all intermediate stages to produce a creation. He then studied the number of these creations over time and found that its growth follows S-curves. Each curve presupposed a final target, a niche size he called a *perceived target*, since competition may prevent one from reaching it. He then proceeded to study hundreds of well-documented artists and scientists. In each case he took the total number of creations known for each of these people, graphed them over time, and determined the S-curve that would best connect these data points. He found that most people died close to having realized their perceived potential. In his words:

> To illustrate further what I mean by the perceived potential, consider the amount of beans a man has in his bag and the amount left when he finally dies. Looking at the cases mentioned here . . . I find that the leftover beans are usually 5 to 10 percent of the total. Apparently when Mozart died at thirty-five years of age, he had already said what he had to say.[1]

The idea is intriguing. Obviously, people's productivity increases and decreases with time. Youngsters cannot produce much because they have to learn first. Old people may become exhausted of ideas, energy, and motivation. It makes intuitive sense that one's productivity goes through a life cycle over one's lifetime, slowing down as it approaches the end. The cumulative productivity—the total number of one's works—could very well look like an S-curve over time. But the possibility of mathematically formulating an individual's peak level of productivity and its inevitable decline before the person dies carries a particular fascination. Marchetti claims to have investigated close to one hundred individuals and found their productivity to proceed along S-curves. I felt compelled to conduct my own investigations. One reason was to check Marchetti's claims; another was to exploit this approach, if confirmed, for my own interests.

To start with, I had to find unambiguous cases of well-documented productivity. It is difficult to quantify productivity. The best-documented individuals are world-renowned artists and scientists. But even then, what constitutes one unit of productivity? By some accounts I found 758 compositions attributed to Mozart, some of them undated, others unfinished or later incorporated into larger works. Granted, composing a minuet at the age of six may not have been a lesser task than a requiem at thirty-five, but should I count every one of the 758 works that have been unearthed and catalogued by zealous Mozart scholars? Would I thus be inflating Mozart's work unfairly? I needed a more objective criterion to decide which ones of Mozart's compositions were worthy of consideration.

Posing this question to musicians, I found that the work had already been done by Austrian musicographer Ludwig Koechel (1800–1877), who studied and catalogued Mozart's works in a way that is universally accepted today. Every "worthy" composition of Mozart carries a K number, according to Koechel's classification. Koechel used numbers all the way to 626, but there are only 568 works indisputably attributed to Mozart today.[2]

MOZART DIES OF OLD AGE

• • •

It is Vienna, 1791. Wolfgang Amadeus Mozart is frantically composing in his sickbed. He has been ill for some time with an undiagnosed disease. He has been composing all along, even more so ever since he fell ill. His illness and rate of composition are going hand in hand. They both intensify. He is only thirty-five years old and has had a brilliant musical career. His wife is deeply worried. She helps in whatever way she can with the music scores and with the sickness, and she wonders about her husband. Is he going to live or die? The doctors are perplexed. Whom can she ask? Who can tell her the future? Exhausted, she slips into a chair and instantly falls asleep.

Back to the future: Geneva, 1987. A VAX-8800 is cranking data through a four-parameter, Chi-squared minimization fit to a logistic function. The data are all Mozart compositions carrying a Koechel number. The fit turns out to be successful. An S-curve is found that passes impressively close to all thirty-one yearly points representing the cumulative number of compositions. There are two little irregularities, however; one on each end.

The irregularity at the low end of the curve caused the program to include an early-missing-data parameter. The reason: better agreement between the curve and the data if eighteen compositions are assumed to be missing during Mozart's earliest years. His first recorded composition was created in 1762, when he was six. However, the curve extrapolates to reach its nominal beginning of 1 percent of the maximum at about 1756, Mozart's birth date. Conclusion: Mozart was composing from the moment he was born, but his first eighteen compositions were never recorded due to the fact that he could neither write nor speak well enough to dictate them to his father.

The second irregularity is at the high end of the curve. The year 1791 shows a large increase in Mozart's productivity. In fact, the

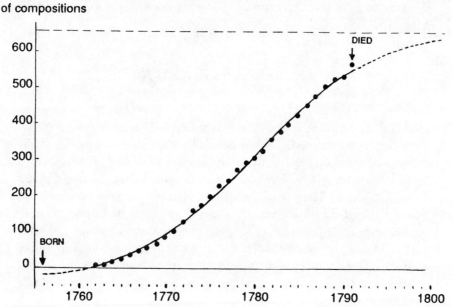

WOLFGANG AMADEUS MOZART (1756–1791)

Cumulative number
of compositions

FIGURE 4.1   The best-fitting S-curve implies 18 compositions "missing" between 1756 and 1762. The nominal beginning of the curve—the 1% level—
points at Mozart's birthday. The nominal end—the 99% level—indicates a
potential of 644 works.

data point is well above the curve, corresponding more to the
productivity projected for the year 1793. What is Mozart trying to
do? His potential is determined as 644 compositions, and with his
last work his creativity will be 91 percent exhausted. Most people
who die of old age have realized 90 percent of their creative potential. There is very little left for Mozart to do. His work in this
world has been practically accomplished. The irregularity at the high
end of his creativity curve indicates a sprint at the finish! What he
has left to do is not enough to help him fight the illness that is
consuming him. MOZART IS DYING OF OLD AGE is the
conclusion flashing on the computer screen.

"Not true!" cries Mrs. Mozart, waking up violently from her
nightmare. "My husband is only thirty-five. He cannot die now.
The world will be deprived of so many musical masterpieces."

• • •

In discussions with musicians I have found that many are not shocked by the idea that Mozart may have exhausted his creative potential at the age of thirty-five. He had already contributed so much in every musical form of the time that he probably could have added little more in another fifty years of calendar time. He himself wrote at the age of twenty-one: "To live until one can no longer contribute anything new to music."[3]

His Dissonant Quartet in C major, K465 (1785), has been cited as evidence for Mozart's possible evolution, had he lived. I consider this an unlikely scenario. The learning curve of music lovers of that time could not accommodate the kind of music that became acceptable more than one hundred years later. Mozart would soon have stopped exploring musical directions that provoked public rejection.

Besides Mozart, I tried to fit S-curves to the works of other personalities from the well-documented world of arts and sciences. Brahms, for example, has 126 musical compositions to his credit. I made a graph of their cumulative number and found an S-curve passing closely to most of the data points approaching a ceiling of 135, Brahms's *perceived potential* (Appendix C, Figure 4.1). That makes Brahms's creativity 93 percent exhausted at the time of his death at the age of sixty-four, which did not seem to deprive him of a chance to realize an appreciable amount of remaining work. The beginning of the curve (the 1 percent of maximum) pointed at 1843, when Brahms was ten years old. This point in time must be considered the beginning of the *pulse of composition* for Brahms. Why, then, didn't he compose until ten years later?

Seeking an answer to that question, I consulted Brahms's biography and found that he had been drawn to music early but for some reason was directed toward piano playing. He distinguished himself as a young pianist. At the age of fourteen he gave public concerts, performing some of the most difficult contemporary pieces. His rare talent for composition manifested itself at the age of twenty, when he suddenly began composing with fervor—three important piano pieces the first year. By 1854 he had nine compositions, as many as there would have been had he spent the previous ten years composing rather than piano playing, but at the more "natural" rate dictated by the pattern he followed so closely until the end of his life.

In retrospect one may say that circumstances made Brahms behave like a pianist between the ages of ten and twenty while his vocation as a composer was brewing inside him, trying to find means of expression. Once it did so, his urge for composition released pent-up energy that made him work feverishly and make up for the time lost.

DID EINSTEIN PUBLISH TOO MUCH?

To further test the hypothesis that one's creativity may grow according to natural laws in a competitive environment, I found myself tabulating Albert Einstein's scientific publications as they are described in a biography published shortly after his death.[4] In Figure 4.2 the data points represent Einstein's cumulative publications, with the S-curve describing the data well. The nominal beginning of the curve (the 1 percent of maximum) points to 1894 when Einstein was fifteen. This would mean that he had no impulse to investigate physics when he was a child. According to the curve, this impulse started when he was a teenager. Still, he produced no publications until the age of twenty-one, probably because nobody would take the thoughts of a mediocre teenage student seriously, let alone publish them.

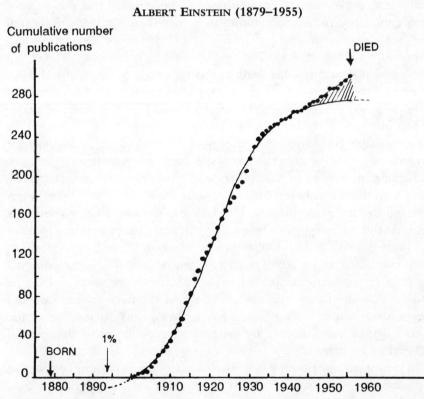

ALBERT EINSTEIN (1879–1955)

FIGURE 4.2   Einstein's cumulative publications and the best-fitting S-curve. The fit indicates 13 publications "missing" between the beginning of the curve, 1894, and Einstein's first publication in 1900. The ceiling is estimated as 279.

The S-curve shown is the one that best fits the data points. And once again the fitting procedure includes the possibility of a parameter representing early missing data, which is determined as equal to thirteen. It means that there are thirteen publications missing for "things" he created between the ages of fifteen and twenty-one that deserved to have been published. Looking into Einstein's biography we find evidence that would support such a conclusion. Even though he did not excel in school performance, by the age of eighteen he had devoured the original writings of the greatest physicists, including James C. Maxwell, Isaac Newton, Gustav R. Kirchhoff, Hermann Helmholtz, and Heinrich R. Hertz.

But the interesting twist of Figure 4.2 is at the other end, toward Einstein's death. His perceived potential as determined by the fit of the S-curve is 279 valid scientific publications. When he died he had 297, overshooting his perceived potential by 8 percent! Looking at the graph we see that during the last eleven years of Einstein's life his publications deviate from the curve, which has practically reached its ceiling around 1945. However, his cumulative works continue growing in a *straight line*, implying a *constant rate* of production, like a machine printing away at a preset rate.

I looked up these late publications. As a rule the original scientific thought in their content is sparse. Some are open letters or honorary addresses at gatherings and social meetings. But most of them are simple translations or old works republished with a new introduction. It would seem that Einstein was overcompensated at the end for what he was deprived of early in his life. Who would dare refuse to publish a document bearing Albert Einstein as its author's name?

CAN WE PREDICT OUR OWN DEATH?

Marchetti pioneered work in correlating the end of life with the end of creativity/productivity by studying selected outstanding personalities from the arts, literature, and science. In a paper titled *Action Curves and Clockwork Geniuses,*[5] he presents curves for scores of personalities. He admits to being biased in choosing cases by his personal interest in the visual arts and the availability of reliable data in his library. The cases he covers include Sandro Botticelli, Domenico Beccafumi, Guido Reni, José Ribera, Sebastiano Ricci, Tintoretto, Francisco de Zurbarán, Johann S. Bach, Wolfgang Amadeus Mozart, William Shakespeare, Ludwig Boltzmann, Alexander von Humboldt, and Alfred J. Lotka. Marchetti concentrates on persons whose actions have been appreciated, studied, and classified. The work done by editors, authors, publishers, and critics contributes to the validation of the data to produce a sample that accurately represents the worthy creations of a genius.

I checked a few of Marchetti's cases (Mozart, Botticelli, Bach, Shakespeare) and ventured to study others on my own. I did not always obtain agreement between a natural-growth curve and the lifework in question. (Some "failures" are discussed in the following section.) My preoccupation from the beginning, however, was that if such an approach is valid, it should not be limited to geniuses. It is based on natural and fundamental principles, and its applicability should be more general. The difficulty with applying it to the ordinary person lies not with first principles but rather with the inability to find the right variable that justly describes one's creativity or productivity. Ordinary people's output has not been measured, appraised, and quantified throughout their lifetimes.

Finding the activity and defining the right variable are the most crucial steps toward deriving one's own natural-growth curve. Activities that depend strongly on physical condition rarely cover the complete life span. A woman's life expectancy extends well beyond the completion of the fertility curve that ends in the early forties. Likewise, the end of a sports career rarely coincides with death. The exhaustion of an artist's or scientist's creativity, however, often corroborates the approaching end. The intellectual domain is therefore better suited for such an analysis than the physical one. Chess matches, for example, are more appropriate than boxing matches. Genuine creativity and productivity in one's area of interest, when systematically quantified and classified, may serve as a good variable. A critical judgment of what constitutes creativity or productivity is essential. It is on that issue that public recognition or rewards are helpful. They provide objective criteria in defining the data set.

It is also very important that the data are complete and bias-free. There must be no gaps due to lost or forgotten events or contributions. There must be no changes in definitions or procedures during data collection. Using multiple sources, that is, two people collecting data on different periods of one's lifetime and combining the results later, may introduce detrimental biases. It is safer to have both investigators cover the whole lifetime and average their findings afterward.

Finally, the fitting procedure must be of the highest standards. Predicting the final ceiling of an S-curve from a limited number of data points may produce very different results among unsophisticated computer programs (presumably no ordinary person will try to do this without a computer). A rigorous Chi-square minimization technique, as described in Appendix A, will ensure the smallest uncertainties. The errors expected from such a fitting procedure will depend only on the errors of the individual data points and the range of the S-curve they cover. Tables for calculating the expected uncertainties are given in Appendix B.

Still, foretelling one's own death may not be so simple. As explained in the next section, a number of complications can result in misleading or unreasonable conclusions. One thing is certain, however; the approach makes intuitive sense. One's creativity and productivity are expected to follow a bell-shaped curve, start low, reach a peak, and then decline toward life's end. One's physical condition follows a similar pattern. The correlation between physical condition and performance is often corroborated in everyday life; for example, when asked to perform, one becomes more resistant to sickness. And in general, maximum productivity coincides with maturity at the highest point in the natural-growth curve of the life cycle.

All this is reasonable and to some extent common knowledge. The new aspect is one of awareness and visualization. The life cycle curve is *symmetric*. Before you rush to quantify your life into data points and try to derive an equation for your own death, just see if you feel you have reached the peak of your productivity in whatever represents your life's activity. If you decide you are somewhere around your peak, comfort yourself with the thought that ahead of you lies a period comparable in length to the one you have had so far.

SEARCHING FOR FAILURES OF THE S-CURVE APPROACH

There are many reasons why a person's creativity or productivity may not follow an S-shaped pattern. The case of a niche-within-a-niche (discussed in Chapter Seven) is one of them, and it can be used to describe, for example, the works of a movie maker who branches out to make television serials. The overall number of his creations is not likely to follow a single smooth curve. It is also possible to have two (or more) niches of activity during a lifetime. A childbearing cycle for a woman may be succeeded by a book-writing one, and one may have a double-barreled life in which a medical career, for example, progresses *in parallel* with an artistic one.

From the moment I became interested in associating mathematical functions to people's lifetimes, I searched as diligently for failures as for successes. In that regard I thought that an unnatural death should abruptly interrupt the evolution of a person's productivity. Consequently, I searched for famous people who died by accident, violent death, or suicide. Soon, however, I came across a new difficulty: What constitutes a *truly unnatural* death? When I tried to construct an S-curve describing Ernest Hemingway's book-writing career, his life's end turned out to look rather natural!

The cumulative number of books Hemingway wrote can be described well by a natural-growth curve whose beginning points at 1909, when he was ten years old. It is no surprise, however, that his first book was not published until fourteen years later; one cannot reasonably expect someone in his teens to publish a book. But as we have seen in previous cases, a blocked creative impulse produces pent-up energy that is released during the early phases of one's career. Hemingway was no exception; he produced intensely the first three years—"unnaturally" so, since his cumulative work grew faster than the curve. But in 1925 the publication of *In Our Time* placed him squarely on the curve, from which he did not significantly deviate until the end (Appendix C, Figure 4.2).

The interesting thing about Hemingway is that the end of his book-writing career is approached smoothly at the age of fifty-two, not unlike Brahms's at sixty-four and Mozart's at thirty-five, both of whom died of natural causes, having accomplished virtually all of their creative potential. Hemingway had exhausted his creative potential. Indeed, biographers have noted that the realization of his waning creative powers was a contributing factor in his suicide. The S-curve depicts Hemingway's death as natural in the sense that it was a natural time for him to die, even though he took his own life.

Suspecting that suicide may not necessarily be different from a natural death, I next examined a purely accidental death: Shelley, who drowned while sailing at the age of thirty. The case seemed complicated because many of his poems were published posthumously in overlapping anthologies. Despite the considerable amount of work involved, I dated each of his poems, an endeavor made possible by the formidable biographical work of Newman Ivey White. The result was rewarding; I was able to graph a steeply rising cumulative number of poems written by Shelley from 1809 onward. When I fitted an S-curve on the data points, I discovered again, as in the case of Hemingway and Brahms, a short early "catching-up" period. Shelley started writing poems at the age of seventeen, while the beginning of the fitted curve points to the age of ten.

What is unique about Shelley's case, however, is that he died leaving his work largely unfinished. The ceiling of the curve indicates a creative potential of twice the number of poems written (Appendix C, Figure 4.3). Apparently, when Shelley died he had said only half of what he would have said, and in his case, his accidental death was unnatural.

I was glad to have done Shelley's curve when I saw Marchetti again at the 1989 International Conference on Diffusion of Technologies and Social Behavior. During a coffee break he explained to the swarm of fans who usually surround him that people die when and only when they become exhausted of creativity, even if their deaths may look like suicides or accidents. "There are no accidents," he proclaimed. I promptly confronted him with Shelley's case. He did not argue. He was willing to allow for exceptions. As I walked away I heard someone say, "Here is a way for life insurance companies to determine whether an accident is real or not before they decide to pay up!"

An Unnatural Suicide Attempt

. . .

Robert Schumann was restless that evening. He gave the impression that it was more than just boredom with the all-too-frequent, musical-social soirée de salon. Around midnight, with two friends still left in his drawing room, he suddenly got up and walked out of the house. His exit was so determined that it prevented his friends from following. It did fill them with worry, however. They were mindful of his chronic mental illness, with its periodic outbursts.

Schumann headed at once for the Rhine and hurled himself into the water. Because of his clothes, he did not sink right away, and this delay gave nearby fishermen who heard the splash a chance to pull him out. In the carriage on the way to the hospital, one of them said, "I think we just saved him." The other one seemed absorbed by another thought. "If he did not succeed in killing himself," he said, "it means he was wrong in trying to do so."

. . .

One hundred and thirty years later evidence is surfacing in support of the concept that Schumann was indeed wrong to attempt suicide at the age of forty-four. He never left that hospital. Even though he died two years later, he never recovered his mental health and did not contribute any more to music. However, the number of his published works kept rising during the eighteen years following his suicide attempt. The numerous musical manuscripts he had left behind appeared before the public at a rate that corresponds to the creative course he might have followed if he had not died. Yet it all happened in his absence!

Figure 4.3 shows the cumulative publication of Schumann's compositions.[6] They oscillate somewhat around the fitted trend. The overall curve is segmented into three periods by his major nervous breakdown in 1845 and his attempted suicide in 1854. Both of these events interfered with the natural evolution of his work, yet one can distinguish a smaller S-curve for each period.

According to the nominal beginning of the process, Schumann must have received his vocation as a composer in 1826, at the age of sixteen. To corroborate this I turned once again to biographies and found that music did not play an important part in Schumann's early life. During his childhood he was not musically inclined but was forced to take piano lessons as a part of his general education. It was not until the age of

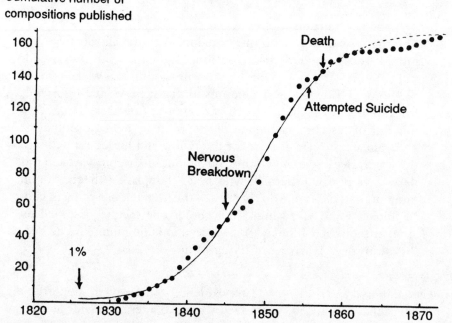

ROBERT SCHUMANN (1810–1856)

FIGURE 4.3 The publication of Schumann's compositions. The fitted curve begins around 1826 and aims at a ceiling of 173. Schumann's publications reached 170, sixteen years after his death.

thirteen that he started enjoying piano practice and taking an interest in choral music. Soon, however, he was sent to law school. He changed universities, trying in vain to study law. It was only when he became twenty that he summoned enough courage to declare openly that he wanted to give up law and devote himself to music. One year later he was publishing musical compositions.

His composing was erratic. It ceased entirely or accelerated quickly following outbreaks of his mental illness. The publication of his works follows a more natural course, however. Society *smoothed out* his irregular composing pattern. His attempted suicide with the ensuing death makes only a small dent in the evolution of his published work. And his publications kept appearing after his death, closely following the curve that had been established to a large extent during his lifetime. As far as the public was concerned, Schumann remained alive, composing, slowing down progressively, and finally stopping in 1872 at the more appropriate age of sixty-two.

Schumann's career is a case where an early forecast with the S-curve approach would have produced erroneous results in spite of the fact that the data may have been validated, well documented, and error-free. By considering publication dates instead of composition dates we have succeeded in obtaining a more regular pattern, but still not regular enough to ensure reliable predictions. Decidedly, there are people whose lives do not follow a natural course and consequently are unpredictable.

A "FERTILITY" TEMPLATE

The creative output of artists and scientists can be quantified by counting their creations as if they were their intellectual children. Would it be instructive to similarly quantify one's real children? The birth rate of American women has been well documented over time. I looked up the data on the birth rate of American women for the year 1987. Plotting the cumulative number, I obtained data points that represent how many children on the average the women of a certain age group had during that year.

Since the numbers were averaged over a large population, statistical fluctuations were small, and I was impressed with how precisely the data points followed the S-shaped growth curve (Appendix C, Figure 4.4). I found that by the age of forty-four the American woman has reached the ceiling of 1.87 children on the average, while at twenty-five she has had only one child. (One can refer to fractions of a child because we are dealing with averages.)

An unexpected insight came from the early phase of the process. To maximize the agreement between data and curve, I let an early-missing-data parameter float to the value of 0.14. The nominal beginning, the 1 percent level of the curve, occurs around ten years of age, where there are no data points. This means that one out of seven American women could have a child "naturally" before the age of twelve. The use of the word "naturally" here has a physiological rather than a social connotation. In any case the general form of the child-bearing natural-growth curve can be considered the template used earlier to describe the intellectual fertility of geniuses. My confidence in this analogy is reinforced with every case where the template works well.

CONTEMPORARY GENIUSES

Having gone this far, I could not resist venturing some forecasts on the careers of celebrated contemporary personalities who are still producing creatively. I considered three Nobel prize winners—two physicists and a writer—and a celebrated movie director.

Burton Richter won the 1976 Nobel prize in physics and is now director of the Stanford Linear Accelerator. He provided me with a complete list of his scientific publications. Carlo Rubbia won the 1986 Nobel prize in physics and is now director of the European research laboratory CERN in Geneva, Switzerland. The data on his scientific publications come from the *Physics Abstracts* (London: the Institute of Electrical Engineers). Gabriel Garcia Marquez received the Nobel prize in literature in 1982. I found his novels catalogued in the library of Columbia University. The career of Federico Fellini, the noted Italian filmmaker, is well documented. To compile a complete list of Fellini's full-length features I consulted three different *Who's Who* publications.

For all four cases I fitted an S-curve on the cumulative number of works and established the level of the ceiling. The procedure yields dates for the 1 percent and the 90 percent levels. The former indicates the moment of conception of the individual's career; the latter points to the time before which the work can be considered largely incomplete and it is unlikely the person will die. The table below summarizes the findings.

Name	Ceiling	Year of 1%	Year of 90%	Born
Richter	340	1954	1991	1931
Rubbia	240	1948	1994	1934
Marquez	54	1964	1994	1928
Fellini	24	1920	2003	1920

We see that the two physicists have comparable careers. They both start young and achieve rough completion of their work at age sixty. Marquez "became" a writer at the age of thirty-six and will continue writing books at least up to age sixty-six. But Fellini breaks the record. It seems that he has at least three more movies to make, the third one at the age of eighty-three!

Fellini's career differs in other ways, too. His curve's nominal beginning points close to his birth date, characterizing him as a born movie maker. Even though he spent much of his early life in movie-related activities, he started directing movies rather late, at the age of thirty-two. By late I mean with respect to his potential. Directing is the job of a boss, and no one starts as a boss. Consequently, for people whose lifework mainly consists of being bosses, there must be an early discrepancy from the natural-growth S-curve. Fellini's curve shows an early catching-up effect not seen in the work of the three Nobel laureates (Appendix C, Figure 4.5). The curve also implies that he should not succumb to disease nor commit suicide before 2003. In fact, all four natural-growth curves indicate that these gifted individuals can look forward to continuing creativity and long lives.

INNOVATION IN COMPUTER MAKING

The next step in my study is to consider the creativity, not of individuals but of a group of people linked together with a common goal, interest, ability, or affiliation; for example, a company or an organization. The individuals now play the role of the cells in a multicellular organism, and their assembly becomes the organism. The population in its entirety, that is, the organization itself, can be seen as an individual. The creativity of such an organization may therefore be expected to follow S-curves in the same way as for individuals.

Alain Debecker and I studied the creative innovation of computer companies, the study that I described in part in Chapter One.[7] We looked at the number of new models introduced by the nine largest computer manufacturers, and by plotting the number of different computer models a company has to its credit, we obtained S-shaped innovation curves.

For Honeywell we played the forecasters' favorite game; namely, to ignore the last five years of history and try to predict it. We fitted an S-curve on the data in the time period 1958 to 1979, and by extrapolating it we obtained excellent agreement with the data in the period 1979 to 1984. At the same time we estimated fourteen models "missing" from the early part of the historical period, probably representing unsuccessful attempts and drawing-board ideas that never materialized.

The most striking feature in the case of several companies was that the fitted curves were approaching a ceiling, thus signifying saturation of innovation and a decreasing rate in the fabrication of differentiated new computer models. This should be of some concern. The computer industry is still young, and innovation reflects on sales volumes. The data used in our study ended in 1985, and the curves drawn at that time implied that companies like Honeywell were in need of new important technological breakthroughs to make possible the development of fresh innovation waves. The ensuing exit of Honeywell from the American computer industry (it was acquired by the French company BULL) may have been symptomatic of the condition pointed out in our analysis.

Another result of our work is that not only new computer models but also new computer manufacturers make their appearance along S-shaped curves. As models fill their niche, manufacturers fill theirs. Both numbers are still increasing today, but as the industry matures these numbers will have to stabilize even if the volume of sales continues to grow. We saw a similar situation earlier this century with the automobile industry. The number of entrepreneurs who launched themselves in car manufacturing grew along a large S-curve to reach a ceiling of about fourteen hundred! This curve was closely followed by another very similar one of business failures reaching almost the same ceiling. The small difference between the two ceilings is the dozen or so car manufacturers who survived and manufacture cars for today's market.

Finally, there is one more observation, already touched upon in Chapter One. Because computer models and computer manufacturers make their appearance in parallel trajectories, their ratio is remarkably stable. A simple graph of the evolution of the number of models as a function of the number of manufacturers reveals a perfectly straight line with a slope equal to five (Appendix C, Figure 4.6). The rate of new models to new manufacturers is constant throughout the whole computer history. It behaves like an *invariant*.

This is more than amusing. It indicates that there is an internal regulation that must be observed in computer making. The number of different models a manufacturer is entitled to over its lifetime is not far from five on the average. How, then, are we to understand prolific giants like IBM, with hundreds of different models to their credit?

To answer this question we should look more closely at what happens around these gigantic citadels of computer know-how. As I suggested earlier, successful models, concepts, and engineers branch off to give rise to small companies, often based around a single idea or person. In other words the mother company produces not only new models but also new companies. There will always be small companies coming and going that never produce five models. But the norm is five, and the more a large manufacturer surpasses this number, the more there will be a need for the appearance of small companies to counterbalance. Therefore, the harmonious way for them to conduct their business, and perhaps the one of least resistance, is to produce new models *and* new companies simultaneously at the ratio of five to one. In this light, individuals who leave their posts to start their own businesses should not be seen as a setback for the employer. These people help keep the balance; without them, pressure may build up somewhere. Their action should be welcomed and to some extent encouraged.

By the same token a good rule for ambitious entrepreneurs who wish to set up new computer companies is to aim either at quick gains to be completed by their fifth model or, if they intend to go beyond this point, to plan to invest in the necessary infrastructure that ensures sufficient technological fertility to give birth not only to new models but to daughter companies as well.

The case studies of this chapter show a clear correlation between creativity/productivity and life span. Yet I would discourage anyone from rushing to predict his or her own death. One reason is that it is difficult to make an evaluation and classification of one's lifework in an impartial way. Another reason is that exceptional circumstances (like those in the life of Schumann) or complications (like the niche-within-a-niche situation) can lead to erroneous predictions. The final reason is that death is not "scheduled" to appear at a particular level of saturation of one's potential. The S-curve approaches its ceiling asymptotically; that is, it needs infinite time to arrive there. The only thing one *can* say is that it is unlikely one may die before the 90 percent level.

There is another argument against indulging in predictions of life's end: It distracts from the more positive benefit that can be derived from the fitted smooth natural-growth curve. A study in detail of the agreement between the curve and the data points can reveal secrets and suggest interpretations of events during one's life. Concerning creativity, the principal value of S-curves may well lie with "predicting" the past.

5

Good Guys and Bad Guys Compete the Same Way

Good and bad are often used to exemplify opposition, not unlike night and day. It is ironic, however, that good and bad are largely subjective notions and something can easily be good and bad at the same time for different people.

Emptying one niche can be seen as filling another one; it is a matter of definition. The number of needles found in a haystack grows with time, but the number of needles remaining to be found declines in a complementary way. In the case of an artist's productivity, the curve that tells how fast his or her output is growing is directly related to the one that describes the rate of exhausting original ideas. The former grows toward a maximum value, while the latter declines toward zero. At all times the sum of the two curves equals a constant, the total number, be it of needles in a haystack or of creations in a lifetime.

For a young child, growing up is always seen as a good thing even though it can be translated to a reduction of the number of days remaining to be lived. Creating a piece of art is never thought of as "one less" among the works remaining to be created. Creating at a reduced rate, on the other hand, carries the mark of deterioration. If I had to give a simple rule for what makes optimists and pessimists, I would say that it is the relative position on the growth life cycle. Before the peak one is optimistic; after the peak the very same events take on a pessimistic hue.

Recently there has been pessimism about the future of the United States. Its power is said to be on the decline, attributable to a variety of social, economic, and political causes, including degenerating morals, a thesis that can provoke chicken-and-egg arguments in response, questioning which is the cause and which the effect.

I would like to propose an observational approach toward decline without trying to explain the reasons that cause it. The only justification for it would be an existential one: Decline must follow because growth preceded. The when and how are details to be estimated through close examination of whatever observables may correlate to the life cycle in question. One such observation is the intellectual strength of a country as it can be measured by the number of its citizens who excel; for example, the number of Nobel prize awards.

We saw in the last chapter that the growth of an individual's achievement is capped. We also saw that companies learn and innovate like individuals. The next level of generalization is to consider a whole country as one organism that grows, competes, and eventually loses to a newcomer. This would imply that the country's collective intellectual achievement must also be capped. I am aware that such a hypothesis may be stretching the analogy, but final judgment should not be passed before a confrontation with the real data.

The United States can be seen as a niche for Nobel awards, and we can map out the growth of the number of prizes it has received over time. The graph at the top of Figure 5.1 shows the cumulative number of Nobel prizes won by Americans since their inception. These results are taken from an article I published in 1988.[1] All disciplines are put together, and each Nobel laureate is counted as one even if he or she shared the prize with others. An S-curve is then fitted to the data.

The agreement between data and curve is good. After determining the function, taking a derivative yields the smooth bell-shaped curve superimposed on the yearly awards data in the bottom of Figure 5.1. This curve depicts the life cycle of the process. At the turn of the century when the awards were instituted, receiving a Nobel prize seemed to be exclusively a European activity. Americans entered the scene in 1907, in small numbers at first, with one award every few years. But Americans gained ground continuously, and by the late 1930s and early 1940s they excelled. From 1958 onward there has not been a single year without American awards. The peak of the smooth curve is around 1978, how-

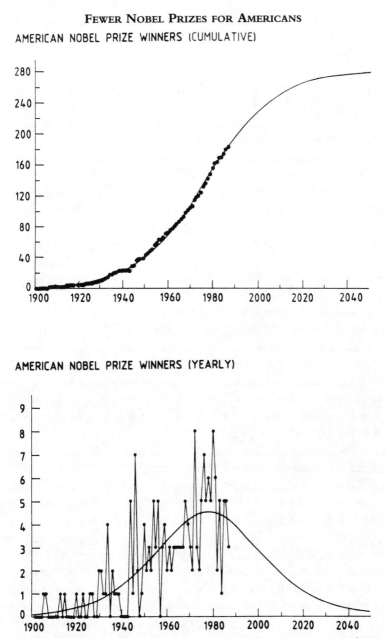

FEWER NOBEL PRIZES FOR AMERICANS

AMERICAN NOBEL PRIZE WINNERS (CUMULATIVE)

AMERICAN NOBEL PRIZE WINNERS (YEARLY)

FIGURE 5.1.   At the top the S-curve fit indicates a "niche" capacity of 283. The real number being 182 makes this niche 64% exhausted by 1987. At the bottom the corresponding life cycle is superimposed on the data.

ever, and it implies entering a declining phase in the 1980s, something that could not have been guessed from the yearly data points, which fluctuate greatly.

The year 1988 saw five American laureates and the year 1989, six. The magnitude of year-to-year fluctuations do not permit the use of these numbers as criteria for judging the forecasts. Furthermore, the uncertainties from the fitting procedure could shift the turnover point in the life cycle by as much as a decade. In any case the average number of yearly American awards by the mid-twenty-first century is projected to be less than one.

The question that comes to mind is, "If Americans are winning fewer Nobel prizes, are others winning more?" To answer this we must consider the distribution of prizes among the competing countries and look for relative trends. Doing so (see Chapter Seven), one realizes that the American decline is, in fact, less alarming than it might at first appear. The American share of Nobel prizes remains rather stable until the year 2000, but a declining absolute number with a *constant* share implies that the total number of Nobel prizes awarded decreases.

Is it reasonable to expect that the awarding of Nobel prizes will weaken and wither with time? Several observations support such a hypothesis. The first is that the yearly number of laureates has been increasing. New categories are added (economics in 1969; mathematics is being considered), and more individuals share prizes more often as time goes on. These inflationary tactics lead to devaluation and suggest that there may be an end of life for Nobel awards.

A second observation is that the average age of Nobel laureates is increasing. It was 54.5 until 1940 and has risen to 57.7 since 1940; if we look at the last ten years only, it has reached 60. Although this may in part reflect the general increase in life expectancy, it is degenerate in nature inasmuch as it associates excellence with the older and weaker rather than the younger and better fit for performance. It is worth noting that age does correlate with performance. At the peaks in award-winning where Americans distinguished themselves brilliantly in the 1930s and the 1950s, the average ages of the recipients were 51.1 and 49.1, respectively.

With this in mind I attempted a fit to the overall number of Nobel awards. I obtained a world potential of 923 laureates over all time, a 56 percent penetration in 1988, and a yearly rate of less than two by the end of the twenty-first century. Thus, a declining annual number of Nobel laureates may at first glance be indicative of an "aging" standard for excellence rather than a loss of intellectual power. To obtain a deeper understanding of American competitiveness, we must look at the performance trend of Americans *relative* to the rest of the world, which we will do in Chapter Seven.

Criminal Careers

Having dealt with outstanding contributors to the benefit of humanity, let us now turn in the opposite direction, to the criminals. We can do that because the careers of many of these individuals have been well documented also and their output can be quantified by the numbers and the dates of their crimes, arrests, or court appearances. Moreover, Cesare Marchetti argues that criminals are not different as far as the *mechanisms* of their behavior is concerned. They simply have "unorthodox" objectives. After all, criminality is socially defined. It is often pointed out that to kill is criminal in peace but heroic in war.

Criminal activities abide dutifully by the laws of natural competition, although competitive strife in this case deals less with rivalry among fellows and more with how to outsmart the police. Police competence is challenged by criminal competence, and the one best fit wins. And because a criminal's career consists of a number of "works," just as in the case of Mozart, one can think again of a presumed potential, a cap or ceiling that may eventually be reached.

The hypothesis is that criminals, like artists and scientists, have a potential for wrongdoing, and their output is regulated according to a final ceiling and a precise schedule. Each has a natural mechanism incorporated that dictates how many crimes will be committed and when. The equation derived from fitting an S-curve to their data reveals this mechanism, and because the equation can be established on a partial set of data, it can be used to predict how many crimes a famous criminal will commit and at what average rate.

It is known that we are genetically programmed in quite a rigorous way, but this form of long-term programming of one's behavior may come as a surprise to those with a firm belief in free will. Before he begins his analysis of criminal careers, Marchetti remarks:

> Human intelligence and free will are the sacred cows of illuministic theology. Free will is already sick from the virulent attacks of Freud, and intelligence is under the menace of astute computer machinery. I will not try to muddle through these loaded and controversial subjects, but take a

very detached and objective view of the situation by forgetting what people think and say and only look at what people do. *Actions will be my observables.*[2]

On the basis of more than a dozen cases in which he established that criminal acts accumulated along natural-growth curves, Marchetti concludes:

> It appears that prison does not have an effect on the global result. The incapacitating periods appear compensated by increased activity once the bird is out of the cage. This hard will to comply with the program and the schedule makes the criminal forecastable. His past activity contains the information to map the future one, in the same sense a segment of the trajectory of a bullet can be used to calculate the previous part and the following part.

Marchetti then proceeds to suggest that limited prison sentences for theft and minor crimes should be abolished. Confinement is ineffective in protecting society because offenders usually "catch up" by accelerated criminality as soon as they are out of prison. He also thinks that detaining people in prisons as corrective action is too expensive and consequently an inefficient way of spending taxpayers' money. A better use of the money would be to compensate the victims, thus treating crime as something like other social misfortunes.

At my first opportunity I argued with Marchetti on this issue. My understanding of competitive growth tells me that natural processes reflect an equilibrium between opposing forces, and that a criminal career would "aim" at a higher plateau had there been no deterrents whatsoever. Fear of a jail sentence certainly helps suppress the expression of criminal impulses. Marchetti agreed with me, but he had an answer ready: "Why not something like old-fashioned thrashings?" he suggested. "They could serve as a deterrent, and they cost far less than prison sentences."

Endless criticisms would certainly be sparked by such arguments. Yet there is a moral to this story, and it should be acceptable to most people. The matching of S-curves to criminal careers—and the ensuing forecasts—can only be done *after* a sizable portion of one's criminality has already been realized. From the point of view of a natural-growth process, this time is too late for effective corrective action. The continued growth of a tree that has already reached a fair fraction of its final height is more or less secured. Measures to really combat criminality effectively should be taken *before* a person establishes him or herself as a criminal.

"Programmed" criminality is not restricted to individuals. In the case of creativity and innovation we have seen similarities between individual and aggregate cases (companies and nations). A company was defined as an organized set of people with a definite common purpose, such as making a profit via manufacturing and selling a product, or providing a service. A criminal organization can be defined similarly even if it has different objectives. In the company's case, the revenue made is one measure of its performance. In a terrorist organization, such as the Italian Red Brigades, the number of victims killed may be taken as a measure of the organization's growth. I have drawn the curve of the Red Brigades by plotting the cumulative number of its victims (Appendix C, Figure 5.1). The data points grow in strict accordance with the natural-growth law. The process is almost complete. The Red Brigades has killed ninety of the ninety-nine people estimated as the ceiling by the fitted curve.

The activities of this terrorist group began in 1974, reached a peak in 1979, and slowed by the early 1980s. The lifetime of the group was nine years. They seem to have disbanded or been annihilated when they realized 90 percent of their "potential." An alternative interpretation of their end is that the group was weakened by "old age" and became more vulnerable to police action. Old age in this context must be interpreted as coming close to the number of operations they had as a final target throughout their existence as a group.

This way of looking at criminality can throw new light on the effectiveness of police operations. Criminal "successes" by the "bad guys" are tantamount to police failures, and complementary curves can be constructed for the "good guys" from the same set of data. The critical interpretation concerns the assignment of cause and effect. Is it the criminals who become weaker or the police who become stronger? In such a tightly knit cat-and-mouse interaction it is difficult to separate cause from effect. A good S-curve description indicates, however, that there is a living entity behind the process observed. Whether it is the criminal organization or the police depends on who is dominating the situation.

KILLER DISEASES

Criminality claims a relatively small percentage of the total number of deaths in society, as do suicides, accidents, and victims of combat action. By far the biggest killers are diseases, particularly those that happen to be prevalent at a given point in time. Today, cardiovascular ailments claim the largest share, close to two-thirds of all deaths, with cancer second at about half that number. A hundred years ago pneumonia and tuberculosis were running ahead of cancer. It seems that diseases take turns in claiming the lion's share of all victims.

The detailed succession between front-running diseases will be looked at more closely in Chapter Seven. In the meantime I want to invoke a notion of *competitive growth* for the number of victims claimed by a disease. The cumulative number of victims of the Red Brigades was seen to follow S-curves, and there was some reason to consider results of criminal activities as social ailments. Why not look at real ailments that way?

Let us visualize diseases as species of microorganisms, the populations of which compete for growth and survival. With the overall number of potential victims being limited, the struggle will cause some diseases to grow and others to fade, a situation similar to different species populations growing to fill the same ecological niche. The extent of the growth of a disease can be quantified by the number of victims it claims. A relative rating is obtained if we express this number as a percentage of all deaths. The percentage of victims claimed by the disease best fitted for survival increases every year, while the unfit disease claims a declining share.

We saw in Chapter One that with respect to deaths from car accidents, society "feels" the risk and keeps tight control over the death rate. Something analogous probably happens in regard to the overall death-rate toll. Looking at the annual death rate per one thousand population in the United States, we find that it was around seventeen at the turn of the century and has been declining steadily to reach values around nine in the late 1980s (see Figure 5.2).[3]

Such a decline is what one expects from all the significant improvements in medicine and living conditions achieved during this century. Important deviations come from events such as wars (for example, 1918) and even then, if spread over a number of years, as was the case with World War II, the deviation washes out.

The important message from Figure 5.2 is the *flattening* of the steep decline. The death rate decreases progressively more slowly, and during the last decade practically not at all. The smooth curve, which is a simple exponential fit to the data, could very well be the final stage of an upside-down S-curve. (Remember, natural growth looks exponential at the beginning and at the end.) No surprises should be expected in projecting this death rate into the future. Society's tolerance level is being approached. From here onward one should expect stability around this

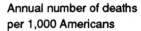

THE DEATH RATE IS NO LONGER DECLINING

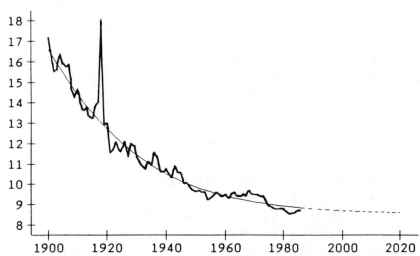

FIGURE 5.2   The death rate in the United States has practically stopped declining. The fitted curve is an exponential or, seen differently, the final part of
an upside-down S-curve. World War I shows up distinctly. World War II is less
visible because it lasted longer and claimed victims in smaller annual percentages
of population.

limiting value, rather than further major declines. If suddenly the death
rate grew significantly, for whatever reason, society's combative efforts
would intensify accordingly until things got back to normal. These efforts
would not go too far, however. The asymptotic level projected in Figure
5.2 of approximately nine deaths per one thousand population per year
seems to be the well-tolerated equilibrium. Expensive emergency action
would stop short of trying to reduce it further.

It is within this rather rigid envelope that the victims of individual
diseases have to be contained. Diseases come and go, and no single one
claims the whole package. Each has its own niche, a fraction of the lot.
These niches are being filled (or emptied) under conditions of natural
competition as diseases elbow one another striving for victims. New and
"gifted" ones have a competitive advantage over old and "tired" ones.
We have often heard that cancer is not a new disease but one that only
became prevalent when others subsided. Like mutants, diseases can be
kept in store indefinitely until favorable conditions bring them out. In

that sense cancer started playing an important role only during this century and consequently is acting as a relatively young entry in the competitive arena of diseases.

If diseases compete among themselves, striving for growth and power in a natural way, the path of filling or emptying their niche should follow an S-curve like the one of species populations and human creativity. The anthropomorphic approach of this hypothesis may be disturbing, but nature is not always delicate; in any case the validity of a hypothesis depends mainly on its verification by real data.

WHEN WILL THERE BE A MIRACLE DRUG FOR AIDS?

To test this hypothesis let us follow the courses of both a "new" and an "old" disease. Diphtheria is one of the latter, and as an old disease it has been fighting a losing battle for survival all through this century. Long before an effective vaccine was developed, diphtheria's share of all deaths was steadily decreasing, a fact that was perhaps known but not fully appreciated.

• • •

Army headquarters, Washington, D.C., 1915. World War I is raging in Europe, but Americans still remain anxious observers. This is another meeting of the Army's top brass, another sequence of heated exchanges among men whose specialty is warfare. Their instinct is warning them of nearing action. The events in Europe challenge these men's training and preparedness. The talk is of numbers: dollars, ships, airplanes, weapons, men, and casualties. They break down the numbers by country, by continent, by army; they project them into the future. They are trying to be as rigorous as they can, to make use of everything they have learned. They want to account for all possibilities, make the best forecasts, the most probable scenarios, leave nothing to chance.

"How much is it going to cost? How many men are we going to lose? How many will the enemy lose?" There are contingencies that seem unforeseeable. "We must anticipate and prepare for all eventualities: bad weather, diseases, earthquakes, you name it. How many men are we going to lose just from ordinary diseases? One out of six people today die from either tuberculosis or diphtheria. These numbers have been declining in recent years, but what if the trends reverse? What if these numbers double or triple from an epidemic among men in action during the critical months?"

"That will not happen," sounds the voice of a psychointellectual maverick attending the meeting. "Trust human nature. When people are called upon to perform, they become more resistant to illness. Soldiers in action develop high resistance, like pregnant women. Under trial they become immune to diseases. They fall ill only when they give up hope."

"I don't know about that theory," snaps a pragmatic military man. "I wish we had a miracle drug for these diseases. What have those scientists been spending all the research money on? Why don't we have a drug to wipe out these diseases? With an effective vaccine in our hands we could instantly eliminate such possibilities for disaster."

• • •

Unknown to them, the men at the meeting had all the elements necessary to make a fairly accurate prediction about one of the diseases that concerned them, diphtheria, but emotions and traditional thinking blocked a more revealing look into the future. The declining jagged line in Figure 5.3 shows the percentage of all deaths attributed to diphtheria in the United States. The data come from the *Statistical Abstract of the United States*.[4]

The years centered around World War I show a wavy structure. The nine-year average is designated with a big dot and falls right on the S-curve fitted to the data points. During the early war years diphtheria claims a particularly low share, while in the years immediately after the war its role becomes excessively important. The situation offers itself to a variety of explanations. Heavy loss of life in combat during the war could result in a reduced share of deaths for the disease. The peak claimed by diphtheria right after the war could be attributed to war-related consequences, such as large numbers of wounded and insufficient supplies and medical care, conditions that may have favored diphtheria more than other diseases for some reason. However, the fact that peak and valley compensate each other rather precisely is evidence for an underlying stability and supports the argument that a concerned population shows resistance to diseases while in danger (that is, World War I) but becomes vulnerable when the danger subsides.

The fit in Figure 5.3 is based only on the data before 1933. The dotted line extended beyond 1933 is the extrapolation of the mathematical

THE PREPROGRAMMED DECLINE OF DIPHTHERIA

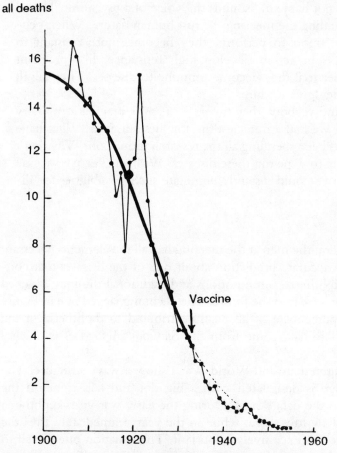

Percentage of
all deaths

FIGURE 5.3   Deaths due to diphtheria in the United States. The big dot is the
9-year average around World War I. The S-curve fit is only on data up to 1933,
the year the vaccine was officially introduced on a large scale.

function determined from the historical window 1900–33. I truncated
history in 1933 because the antidiphtheric vaccine was officially introduced on a large scale that year. In a country like the United States, with
excellent means of communication and an efficient medical system, one
would expect that once a vaccine becomes available for a bacterial disease, there will be no more victims.

Surprisingly, the historical data after 1933 coincide with the extrapolation of the fitted curve determined from the data before 1933. It means

the disease continued to decline according to a "prescription" established *before* the vaccine. If anything, the data points seem to fluctuate less in more recent years, making the agreement between the forecasted trajectory and the actual course more striking. At face value this would imply that the miracle drug in this case had no real effect!

I found this conclusion so disturbing that I went to the library to read about the antidiphtheric vaccine. It was not an inspired discovery. Researchers had been working on it for decades with progressive success in both Europe and the United States. Early, less efficient versions of the vaccine had been employed on large samples of the population. The more I read, the clearer the emerging conclusion became. The development of the vaccine had itself gone through a natural-growth process over many years, and what we see in Figure 5.3 is the combined evolution of disease *and* vaccine.

Medicine was not the disease's only enemy. Diphtheria's decline must also be attributed to improvements in living conditions, diet, hygiene, or simply education. Moreover, fast growing new diseases, such as cancer, claimed increasing shares of the constraining overall death toll. The fully developed vaccine helped regulate the exit of diphtheria rather than eliminate it instantly. Individuals would still die from it, but epidemic outbursts where whole villages were decimated could now be prevented. The vaccine's history is intertwined with the disease's phasing out. The progressive realization of the antidiphtheric vaccine can be seen both as cause and effect of the phasing out process.

A situation similar to diphtheria is found with tuberculosis. The medical breakthroughs that had an impact on this disease include the development of antibiotics and, of course, the B.C.G. (Bacillus-Calmette-Guérin) vaccine. The first human experiment with the tuberculosis vaccine was performed in 1921 in France, but ten years later its safety was still disputed by experts such as the distinguished American bacteriologist S. A. Petroff.[5] Reliable curative antituberculosis drugs became available only in the mid-1950s. Which one should be labeled the "miracle" drug is not obvious. What is clear for tuberculosis, however, just as it was for diphtheria, is that the disease has been continuously losing ground to other diseases since 1900. Repeating the diphtheria exercise I graphed the percentage of all annual deaths attributed to tuberculosis from 1900 onward. Then I fitted a growth curve to the historical window 1900 to 1931, before penicillin, antibiotics, vaccine, and curative drugs. The projection of this curve agreed impressively well with the data of a much later period, the post-1955 years. Such an agreement must be attributed to a lucky coincidence. Several similar S-curves could have been accommodated within the uncertainties of the fitting procedure, but all of them would display the same general pattern. The data points *after* effective medication became available trace the same pattern with exceptional regularity (Appendix C, Figure 5.2).

Instead of the pronounced deviation at the time of World War I seen in the case of diphtheria, the tuberculosis data showed a broad peak around 1946, with an increase from 1936 to 1946, the decade leading up to and including World War II. Such a deviation might be explained by the poverty and hardships of the Depression, conditions that could result in a comeback for tuberculosis.

I have seen occasional deviations on the shares of other diseases phasing out along declining S-shaped trajectories: typhoid, whooping cough, scarlet fever, gastroenteritis, and pneumonia. Local deviations can sometimes be due to improper classification. For example, data on propeller and jet airplanes, both of which have competed with ships in transatlantic passenger transport, when lumped together do not produce a smooth growth curve for airplanes. Similarly, the lumping together of the Spanish flu—which may not have been a flu at all—with pneumonia produces a noticeable upward deviation at the end of World War I.[6]

With time, however, such deviations from the overall trend are eventually reabsorbed without leaving a long-term trace. The image that emerges rather reliably from the cases studied so far is that a phasing out disease empties its niche—its share of all deaths—along an S-shaped trajectory. The essential message can be summed up as follows:

- Winning the battle against a disease is a process that proceeds naturally. From historical data on the first half of the process one can forecast the second half.
- A disease starts phasing out well before effective medication becomes perfected and widespread.

In this light the hope for a miracle drug as the only way of combatting AIDS seems to be a naive expression of wishful thinking. If these hypotheses are true, discovering a vaccine or other effective cure for AIDS is unlikely before the numbers show a relative decline in AIDS victims. This conclusion should not be cause for alarm. The number of AIDS victims in the United States is far from declining, but the growth in the number of new cases is slowing down. During the last decade AIDS claimed a progressively bigger share of the total number of deaths every year. Its "niche" however, seems to be far smaller than that feared by most people. When I plotted the evolution of the disease in terms of its share of all deaths, I obtained the mirror image of diphtheria.

The data on deaths from AIDS came from "The AIDS Surveillance Report" issued by the U. S. Department of Health. They, as well as many other agencies worldwide, have been graphing and monitoring the spread of the disease by looking at headline statistics only. What is new in my analysis of AIDS is that by considering its *share of all deaths* we are looking at the relative strength of the disease, thus revealing the competitive struggle between AIDS and other diseases.

The S-curve I fitted on the data indicates a growth process that is almost complete by the end of 1988. (I left out the two semesters of 1989 to avoid biases from delays in reporting.) The ceiling for the disease's growth is projected to be 0.95 and should be reached in the early 1990s (Appendix C, Figure 5.3). In other words a place seems to be reserved for AIDS in American society just below the level of 1 percent of all deaths, as if there were other, more important causes of death. Oscillations in the number of AIDS victims may follow later on in this decade, and there is always the possibility of large chaotic deviations above and below the 0.95 percent level (see the discussion on chaos in Chapter Ten). But for the time being AIDS is approaching a ceiling.

There seems to be a mechanism that is limiting AIDS in a natural way. This mechanism may reflect the control exercised by American society through collective concern. With or without a miracle drug the disease is not likely to spread further and indeed is likely to begin declining in the future. Eventually there will be effective medication for it. Those who predict imminent doom without the discovery of a miracle drug fail to take into account the natural competitive mechanisms that regulate the split of the overall death rate among the different causes of death, safeguarding all along an optimum survival for society.

6

A Hard Fact of Life

Natural growth in competition in its simplest form is a process in which one or more "species" strive to increase their numbers in a "niche" of finite resources. Depending on whether or not the species is successful over time, its population will trace an ascending or a descending S-curve. In a niche filled to capacity, one species population can increase only to the extent that another one decreases. Thus occurs a process of *substitution,* and to the extent that the conditions of competition are natural, the transition from occupancy by the old to occupancy by the new should follow the familiar S-curve of a natural-growth process.

The first connection between competitive substitutions and S-curves was done by J. C. Fisher and R. H. Pry in a celebrated article published in 1971. It became a classic in studies of the diffusion of technological change. They wrote as follows:

> If one admits that man has few broad basic needs to be satisfied—food, clothing, shelter, transportation, communication, education, and the like—then it follows that technological evolution consists mainly of substituting a new form of satisfaction for the old one. Thus as technology advances, we may successively substitute coal for wood, hydrocarbons for coal, and nuclear fuel for fossil fuel in the production of energy. In war we may substitute guns for arrows, or tanks for horses. Even in a much more

106

narrow and confined framework, substitutions are constantly encountered. For example, we substitute water-based paints for oil-based paints, detergents for soap, and plastic floors for wood floors in houses.[1]

They went on to explain that, depending on the time scale of the substitution, the process may seem evolutionary or revolutionary. Regardless of the pace of change, however, the end result is to perform an existing function or satisfy an ongoing need differently. The function or need rarely undergoes radical change.

A similar process occurs in nature, with competition embedded in the fact that the new way is vying with the old way to satisfy the same need. And the eventual winner will be, in the Darwinian formulation, the one better fit for survival. This unpalatable conclusion may seem unfair to the aging, wise, and experienced members of human society, but in practice it is not always easy for the young to take over from the old. Experience and wisdom do provide a competitive edge, and the substitution process proceeds only to the extent that youth fortifies itself progressively with knowledge and understanding. It is the combination of the required proportions of energy, fitness, experience, and wisdom that will determine the rate of competitive substitutions. If one's overall rating is low due to a lack in one area, someone else with a better score will gain ground, but only proportionally to his or her competitive advantage.

The opposite substitution, cases in which the old replace the young, is possible but rare. It is sometimes encountered in crisis situations where age and experience are more important to survival than youth and energy. But independently of who is substituting for whom, and in spite of the harshness ingrained in competitive substitutions, one can say that the process deserves to be called natural.

Let us look at some impersonal examples. A classical natural substitution of technologies was the replacement of horses by cars as means of personal transportation. When automobiles first appeared, they offered a radically different alternative to traveling on horseback. The speed and cost, however, remained roughly the same. In *Megamistakes: Forecasting and the Myth of Rapid Technological Change,* Steven Schnaars considers a cost-benefit analysis as one of the three ways to avoid wrong forecasts.[2] In this case such an analysis would have predicted a poor future for cars.

Ironically, one of the early advantages attributed to the automobile was its nonpolluting operation. In large cities, removing horse excrement from the streets was becoming a serious task, and projected transport needs presented this problem as insurmountable. For this and mostly for other more deeply rooted reasons, cars started replacing horses rapidly and diffused in society as a popular means of transportation.

We can look at this substitution process in detail by focusing on the early automobile period 1900–1930. We consider again the relative amounts only, that is, the percentage of cars and horses of the total number of transport "units" (horses plus cars). Before 1900, horses filled 100 percent of the personal transport niche. As the percentage of cars grew, the share of horses declined, since the sum had to equal 100 percent. The data in Figure 6.1 show only nonfarming horses and mules.

These trajectories are seen to follow complementary S-curves. In 1915

THE SUBSTITUTION OF CARS FOR HORSES IN PERSONAL TRANSPORTATION

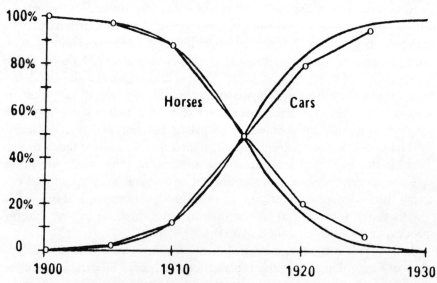

FIGURE 6.1   The data points connected by straight lines represent percentages of the total number of transportation units, namely, cars plus nonfarming horses and mules. The S-curves are not fits to the data points; they are simply drawn to illustrate the idealized substitution path. The sum of respective ascending and descending percentages equals 100 at any given time.*

*Adapted from Cesare Marchetti, "Infrastructures for Movement," *Technological Forecasting and Social Change,* vol. 32, no. 4 (1987): 373–93. Copyright 1986 by Elsevier Science Publishing Co., Inc. Reprinted by permission of the publisher. Credit for the original graph must be given to Nebosja Nakicenovic, "The Automobile Road to Technological Change: Diffusion of the Automobile as a Process of Technological Substitution," *Technological Forecasting and Social Change,* vol. 29: 309–40.

there are an equal number of horses and cars on the streets, and by 1925 the substitution is more than 90 percent complete. It is interesting to notice that the S-curve drawn in an idealized way (the curves here are not fits to the data) goes to 100 percent for cars after 1930, while the data points seem to aim at a lower ceiling. This may be related to the fact that a certain number of horses were not replaced by cars. They are probably the horses found today in leisure riding and horse racing.

To take another example, more recently steam locomotives were replaced by diesel or electric ones in most parts of the world. The old "species" declined in a way reminiscent of diphtheria and tuberculosis. In his book, *The Rise and Fall of Infrastructures,* Arnulf Grubler shows that the percentages of steam locomotives in the United States and the U.S.S.R. declined just as the percentage of horses had done earlier. The raw data outline an S-curve so clearly that there is no need to fit them with a curve (Appendix C, Figure 6.1).

The data also show that the Russian and American steam engine decline curves are ten years apart but strictly parallel. This similarity is deceptive, however, because of another substitution that was taking place. These two countries not only differ geographically and culturally but their dependence on railways is fundamentally different. By 1950, midway through the phasing out of steam in the United States, transportation by railway seemed to have already yielded to road (highway) transport, with seven times more roadway length than railway track (see Chapter Seven). In contrast, by 1960, halfway through the steam era in the U.S.S.R., railways still flourished with tracks totaling half the length of all paved roads put together.

Examples of obsolescence such as horses and steam locomotives illustrate the inevitable takeover by newcomers possessing competitive advantages. There are always two complementary trajectories in one-to-one substitutions, one for the loser and one for the winner. They indicate the shares, the relative positions, of the contenders. The ceiling is by definition 100 percent, a fact that makes the determination of these curves easier and more reliable than the ones used earlier in this book, in which the ceiling had to be ascertained.

By looking at shares we focus on competition and demonstrate an advantage whose origin is deeply rooted, as if it were genetic. This description thus becomes free of external influences: economy, politics, earthquakes, and seasonal effects such as vacations and holidays. In the case of locomotives the popularity of trains did not affect the substitution process. In the case of horses and cars, World War I had no impact!

Another advantage of focusing on relative positions in substitution processes is that in a fast-growing market the natural character of the process (the shape S) may be hidden under the absolute numbers that are increasing for *both* competitors. The fact that one competitor is growing faster *and* that the percentage trajectory follows a natural path is evidence that the other competitor is phasing out. In Figure 6.2 we see the absolute numbers of horses and cars in the United States since 1850. During the first decade of this century the number of horses continued to increase as it had in the past. The number of cars increased even more rapidly, however. The substitution graph of Figure 6.1 reveals an indisputable decline for the horses' share during the decade in question.

THE OVERALL PERSONAL "VEHICLE" MARKET KEPT INCREASING EVEN AS THE USE OF HORSES DECLINED

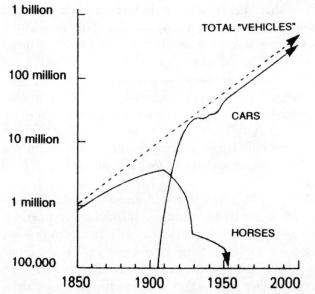

FIGURE 6.2   The numbers of nonfarming horses (including mules) and cars in the United States. The vertical scale is logarithmic to accommodate the many-fold growth. The sum of horses plus cars grew practically unaffected across the substitution period.*

*Adapted from a graph by Nebosja Nakicenovic in "The Automobile Road to Technological Change: Diffusion of the Automobile as a Process of Technological Substitution," *Technological Forecasting and Social Change,* vol. 29: 309–40. Copyright 1986 by Elsevier Science Publishing Co., Inc. Adapted by permission of the publisher.

Playing the devil's advocate, one may try to use this way of thinking to prove that given any two competitors, one of them must necessarily be phasing out. Since they are bound to grow at different rates, one will always be "losing" relative to the other. The fallacy of this approach is that two competitors chosen at random most probably do not constitute a niche. A natural one-to-one substitution is expected whenever there is direct transfer from one to the other with no third parties involved. The niche needs to be carefully defined. For example, do the two computer makers IBM and Digital together constitute one niche? Should we expect substitution of one by the other, or should they be looked at only as two out of many members of the bigger computer market? One evidence for such a microniche would be that the two shares follow S-shaped trajectories over time. Other evidence may be to know independently that in a certain geographical or market segment these two are the only companies that offer products. In the latter case the geographical or market segment constitutes a niche, but whether or not a substitution process is taking place in a *natural* way will depend on how closely the market share evolution resembles an S-shaped pattern. This pattern can be made to appear as a straight line when viewed through the appropriate "eyepiece."

STRAIGHT LINES IN NATURE

A popular belief claims that there are no straight lines in nature. Precise mathematical shapes such as circles, flat surfaces, right angles, and ellipses are not encountered during a stroll through the woods. These shapes are thought of as human inventions; therefore, they are stamped "unnatural." Rudolf Steiner, a twentieth-century Western guru with a significant following in Central Europe, built his philosophy around "The Natural." Among other things he did, he made sure that an enormous concrete building was erected—itself a monument to his theories—that featured practically no right angles.[3]

Could a natural phenomenon produce a straight line? Undoubtedly yes; I can think of many examples, but I want to concentrate on natural growth in competition, which so far has been described as following an S-shape. The logistic function used in the one-to-one natural substitutions is such that if one divides the number of the "new" by the number of the "old" at any given time and plots this ratio on a logarithmic vertical scale, one gets a *straight* line. It is a mathematical transformation. The interest in doing this is to eliminate the need for computers and sophisticated fitting procedures when searching for "naturalness" in substitution processes.

Consider the following hypothetical case. You are concerned about the new hamburger stand that opened recently across the street from yours. You worry that its guarantee of home delivery in less than ten minutes will not only cut into your business but put you out of it altogether. People may just decide to save the gas and order their sandwiches from wherever they happen to be. To make it easy, let us also suppose that the stand across the street has set up a public counting device which advertises the number of hamburgers it has sold so far.

It is easy to find out if your competitor's way is going to be the way of the future. Here is a recipe. Start your own count of how many hamburgers you have sold since the competition appeared and track the ratio—your competitor's number divided by your number. Then run to a stationery store and buy some semilogarithmic graph paper on which you will plot this ratio every day. By the third day you should be able to see how this ratio is lining up with the previous two. Several weeks later you will have enough data points to discern the trend. If the data points form a straight line (within the small daily fluctuations, of course), you may want to start looking around for another business. If, on the other hand, the trend is flattening or the pattern is such that no overall straight line is discernible, then chances are the guys across the street are not here to stay, or if they are, they will not take all the business.

Looking at a natural substitution process in this way reveals a straight line. The straighter the line, the more natural the process. The longer the line, the more confidence one can have in extrapolating it further. If the ratio of new to old has not yet reached 0.1, it may be too early to talk of a substitution; a natural substitution becomes "respectable" only after it has demonstrated that it has survived "infant mortality."

A striking example of this process is shown in Figure 6.3, taken from the Fisher and Pry article cited previously. It shows the substitution of detergent for soap in the United States and Japan as it is reflected in the percentage rise in the total market share. (The corresponding decline for soap is not shown.)

Detergents entered the market in the United States right after World War II and arrived in Japan ten years later. In both countries the substitution of detergent for soap has now reached more than 90 percent of the market niche. This does not mean that all soap became extinct. Detergent replaced laundry soap only. Beauty soap has been thriving all along in its

DETERGENTS SUBSTITUTING FOR SOAP

**Detergents as a percentage
of total market**

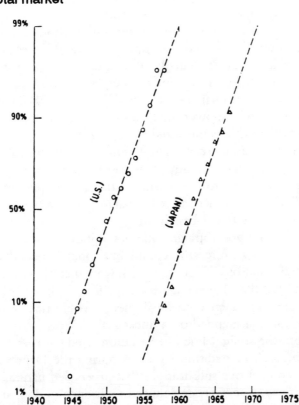

FIGURE 6.3   The data points represent the ratio of detergents to soap consumed annually in the United States and Japan. The vertical scale is logarithmic, which makes the fitted S-curves (intermittent lines) appear straight. The numbers at the scale gradations do not refer to values of the ratio, however, but to market shares instead. We can say that a graph constructed in this way makes use of a *logistic* scale. We see the rise of detergents; the complementary decline of soap is not shown.*

*Adapted from a graph by J. C. Fisher and R. H. Pry in "A Simple Substitution Model of Technological Change," *Technological Forecasting and Social Change,* vol. 3, no. 1 (1971): 75–88. Copyright 1971 by Elsevier Science Publishing Co., Inc. Reprinted by permission of the publisher.

own separate niche. It is also interesting that in spite of the ten-year delay and the enormous cultural differences, the rates of substitution in the two countries seem identical, as reflected in the inclination of the lines. One wonders if there is any significance in this similarity.

Cases of competitive one-to-one natural substitutions abound. Nebojsa Nakicenovic has studied technological substitutions and has come up with many examples that fit the straight-line description. The longest one is between two broad categories of primary energy sources: traditional and commercial. In the traditional energy sources category Nakicenovic put fuel wood, water and wind power (mills, waterfalls, and so forth), and the output from work animals as measured by the energy content of the food consumed. In the commercial energy sources he included coal, crude oil, natural gas, and nuclear energy. This classification is particularly interesting because, in addition to the cultural distinctions implied in the terms "traditional" and "commercial," it separates renewable energy sources from nonrenewable (fossil-based) ones.

Nakicenovic obtained a good straight-line description for the substitution of traditional energy sources by commercial ones (Appendix C, Figure 6.2). Ninety percent of U.S. energy needs were met by renewable sources in 1850, but less than 1 percent are today. This decline has been a natural process. The dependence on fossil energy would trace out an almost complete S-curve if graphed on a linear scale. Occasional short-term deviations from the straight line can be rationalized—for example, the short-term comeback of traditional sources during the Depression.

There are many competitive substitutions that proceeded naturally in the automotive industry. Cars were introduced with open bodies, and it took close to thirty years for 90 percent of them to be enclosed. More recently and at a faster pace—ten years—disk brakes replaced drum brakes, and radial tires replaced diagonal ones. Similarly, power steering and unleaded gasoline are in the process of taking over from their predecessors.

Natural substitutions are responsible for most popular trends and would trace a straight line on a logarithmic scale. Consider the American workforce, for example. Among the megatrends of today is the growth in the information content of jobs and in the increasing role of women. With data from the Bureau of Labor Statistics and through my analysis I found straight lines for the historical evolution of the respective percentages. The extrapolations of these lines quantify the trends until the end of the century. Projections obtained in this way should be rather reliable because they follow the natural-growth path of the process (Appendix C, Figure 6.3).

Noninformation workers are manual laborers in service occupations, manufacturing and repair, construction, farming, forestry, and fishing. The percentage of these workers has been steadily decreasing. The 50 percent point was reached back in 1973. By the year 2000 it is reasonable to expect less than 40 percent of the workforce in this category. (It should be noted, however, that such categorization underestimates the amount of actual "information work" since manual laborers do spend a significant part of their time at tasks requiring information handling.)

The fact that women have been taking an increasingly important role in the workforce is common knowledge. What has been less publicized is the *kind* of work they do. Among the information workers 56 percent were women in 1991, largely in clerical, secretarial, and administrative positions. If one looks at executives, the picture is different. I plotted the percentage of executives who are women and again obtained a rising straight line. Women's role as executives has been steadily growing but will reach 50 percent in a natural way only around the year 2000.

A natural substitution process, in the absence of "unnatural" interference, should proceed to completion. In my graph I compiled data according to which women executives progressively replaced men during a twenty-one-year period, although the trend had probably been established well before that time. Under the persistence of similar conditions, women should continue the substitution process until eventually more executives will be women than men.

Could women really dominate the executive scene? Why not, one may argue? Not so long ago an imbalance of virtually 100 percent male executives did not stop society from functioning. An S-curve extrapolation of a trend over a period comparable to the historical window is defensible. Such an extrapolation would raise the share of women among executives to 60 percent by the year 2010. But before we extrapolate further, we must take into account some other considerations.

Sometimes substitutions do not reach 100 percent because new contenders enter the competitive scene. For example, cancer replaces cardiovascular diseases, but it will never claim 100 percent of all deaths because AIDS or another new disease will grow to claim a fraction. Oil replaced coal along a natural trajectory for a while (see Chapter Seven), but well before a 100 percent substitution was reached, natural gas began replacing oil. In the case of women replacing men, however, we cannot reason the same way because we do not expect another sex to start taking over from women. Yet a fundamental change may still occur.

The definition of an executive may evolve, for example. Recent directions in management show moves away from the traditional hierarchical structure. Companies are experimenting with removing power, responsibility, and decision making from the top and delegating it to unit managers and the people close to the customer. The niche of executives, as we have defined it so far, may cease to exist before it becomes dominated by women. Alternatively, women's advance may be prevented from following its natural path to the end. The data points for the last three years happen to be somewhat below the curve (Appendix C, Figure 6.3). If this is not an "innocent" fluctuation, it might be early evidence of a quota assigned by some social mechanism.

THE CULTURAL RESISTANCE TO INNOVATION

The rate of substitution between technologies is not only a consequence of technological progress but also reflects society's rate of acceptance of the innovation in question. We will see in the next chapter that it took more than one hundred years for sailing ships to replace steamships. This time span is much longer than the lifetime of the ships themselves. Sailing ships did not persist on the seas because they were made to last a long time or because it was too costly to convert shipyards to outfit boats with steam engines. There is a certain resistance to innovation, which is controlled by *cultural* forces. Rational motivations such as costs and investments play a secondary role.

The growth of the American railway network required not only huge investments and construction efforts but also a battle against the strong social opposition to the diffusion of a new technology, as expressed in a poster shown in Figure 6.4.

Another innovation vehemently opposed is the fluoridation of drinking water to prevent tooth decay. Historian Donald R. McNeil has summed up the situation in an article titled "America's Longest War: The Fight Over Fluoridation, 1950–19??[4] He points out that after forty years, half of the American population does not have fluoridated water. The opposition was varied and surprising. Even the Church was mobilized against fluoridation. Among the arguments used by opponents has been that it increases communism, atheism, and mongolism! Water supplies in the United States are routinely chlorinated today, but that idea was also fiercely opposed when first suggested.

Opposition to technological innovations reached violent and criminal levels in England in 1830 when farming machines were introduced. The increased productivity of threshing machines jeopardized the jobs of farm

A VERY VOCAL OPPOSITION

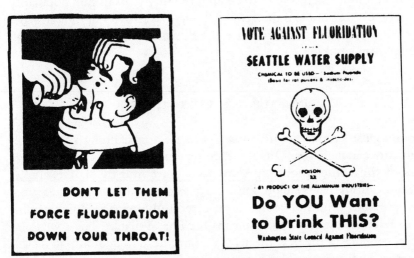

FIGURE 6.4 These posters exemplify cultural resistance to technological change. The railway poster was provided courtesy of Metro-North Commuter Railroad, New York. The two posters at the bottom are from Cesare Marchetti, "On Society and Nuclear Energy," report EUR 12675 EN, 1990. Reprinted by permission of the Commission of the European Communities, Luxemburg.

laborers, who reacted by trying to stop the diffusion of the innovation and tried to win public opinion and the sympathy of the Church. Opponents first manifested sporadically by destroying farming equipment, setting farms ablaze, sabotaging firefighting efforts, and receiving light sentences for such crimes. But troubles culminated in November 1830 with a wave of concentrated attacks on threshing machines. E. J. Hobsbawm and G. Rude have collected detailed data on the destruction of 250 machines in a period of one month.[5] This violent expression of discontent displayed the hallmark of a "natural" evolution: The cumulative number of machines destroyed traces a complete S-curve in one month (Appendix C, Figure 6.4). Similar pictures would probably describe other means of resistance that also slowed the diffusion of this innovation.

The introduction of farming machines capable of increasing productivity is typical of a new technology disturbing an equilibrium established around an old technology. The part of the population whose income is threatened will fight against it for a while, retarding the rate of diffusion. But sociocultural resistance to change was not always manifested by violent opposition and criticism. Sometimes its only manifestation was the introduction of extended delays in what should normally be rapidly accepted. Innovations of indisputable benefit requiring no capital investment or economic hardship have taken up to fifty years to gain popularity in society. One example is the general theory of relativity; it took decades for physicists to accept it. Another is the slowness with which Americans are giving up smoking.

UNNATURAL SUBSTITUTIONS

Commonsense arguments may suggest that a forthcoming substitution should proceed naturally. One can then reasonably expect the percentage shares of the competing entities to behave accordingly. For example, communications that were predominantly exchanged through letter writing in previous centuries could be expected to be replaced by telegrams and/or telephone calls as these means became available. The speed and the convenience of the new means of communication demonstrated clear competitive advantages. This turned out to be the case for telephone calls but not for telegrams.

Arnulf Grubler demonstrated that substitution of telephone calls for letters in France since the turn of the century proceeded along straight lines (Appendix C, Figure 6.5). The data he graphed were the percentages of all messages exchanged by telephone or letter during one year. Despite fluctuations, some of them understandably due to the war (people wrote letters during the war but rushed to telephones as soon as the war was over), the overall trend seems to indicate a natural process of substitution, starting sometime before the beginning of this century and reaching 70 percent by 1980. Telegrams, and later telexes, were not considered because they never gained more than a fraction of a percent of the total messages market, Grubler said. He did not look at facsimile transmissions for lack of data, but it is safe to say that the telephone had by far the biggest impact on communications.

What resisted substitution the most was face-to-face contact in communications. When the telephone was introduced, people hoped it would replace some of the necessity for travel. Recently, more sophisticated means of electronic communication have become available—teleconferences and image telephones—that could also be expected to replace business trips, particularly during times of cost controls. Such technology has undoubtedly eliminated a certain amount of legwork. The need for personal contact seems to stubbornly resist replacement, however. The effectiveness of face-to-face dealings is recognized by business persons today as much as it was by the emissaries of antiquity.

Serious doubts are raised as to whether technological means will ever satisfy the social need for communicating through personal contact. It is difficult to compare quantitatively the amount of information exchanged through personal contact with that exchanged through technological means. The informational content of body language, for example, cannot easily be measured in units, like phone calls or pieces of mail. But the insistence with which executives and politicians cling to their costly missions suggest that such a substitution may be "unnatural" or, at best, a slow one, much slower than the letter-to-telephone substitution.

Another reason for which an a priori reasonable substitution may not display the expected trajectory is that the variable is not properly defined. The example of detergent substituting for soap mentioned earlier would not have followed a straight line had Fisher and Pry taken *all* soap. They carefully considered laundry soap only, leaving cosmetic soap aside. Complications can also arise in substitutions that look "unnatural" in cases of a niche-within-a-niche or a niche-beyond-a-niche. In both cases two different S-curves must be considered for the appropriate time periods. In the straight-line representation such cases show a broken line made out of two straight sections.

When close scrutiny does not eliminate irregularities, it means that there is something unnatural after all. A substitution may show local deviations from a straight-line pattern, which can be due to exceptional temporary phenomena. But such anomalies are soon reabsorbed, and the process regains a more natural course. As an example of this, Fisher and Pry point out the replacement of natural by synthetic rubber during the war years. In the 1930s synthetic rubber was slowly making an appearance in the American market as an inferior alternative to natural rubber, which was imported in large quantities from foreign sources. During the early stages of World War II, imports of natural rubber were largely cut off, and at the same time demand for rubber increased considerably. A large national effort during these years resulted in improving the quality and reducing the production costs of synthetic rubber.

Fisher and Pry have shown that synthetic replaced natural rubber at an accelerated rate during the war years (Appendix C, Figure 6.6). But as soon as the war ended, foreign sources of natural rubber became available again, and the substitution rate dropped. From then on the substitution process continued at a rate similar to other replacements, such as margarine for butter and synthetic for natural fibers. The deviation, caused by the necessities of the war, disappeared leaving no trace when life returned to normal.

The cases presented in this section depict deviations from the description of the natural substitution model. In the work of economists and meteorologists such a situation—evidence *against* a model—would have resulted in modifications of the model itself. With the process of natural growth in competition, however, the theory is fundamental; deviations, whenever they occur, must be explained in the way the model is used rather than raise questions about its validity.

THE SWEDES ARE COMING

• • •

October 1956 in Hungary. Russian tanks are rumbling on the streets of Budapest, crushing the popular uprising. There are bloody clashes when students, workers, and Hungarian soldiers oppose the Red Army. Casualties run high; leaders fall, and so do hopes for democracy. Emotions heat up; the young do not want to abandon their ideals. Clandestine meetings take place where ambitions are transformed into daydreaming.

"If only a superpower would take pity on us and come down on the Russians with mighty weapons to punish them for this wrong-doing!"

"You think the Americans might do it?"

· · ·

The Americans did not do it. The whole world was alarmed but watched passively as the rebellion was ruthlessly repressed and 170,000 Hungarians fled their country. It was only later that a bee appeared to have stung the bear.

While I was exploring the distribution of Nobel prizes among different countries, I found substitution processes taking place. Some trends seemed familiar; for example, Americans progressively gained over Europeans during the first half of the century. But one substitution, in spite of its natural appearance, did not make sense.

There has been a twenty-five-year-long battle between Swedes and Russians for Nobel prizes. I focused on these two countries when I noticed that from 1957 to 1982 the Russian share of prizes declined sharply while the Swedish share showed a clear rise. The slopes being complementary made me suspect a one-to-one substitution in a local microniche.

On further investigation I found that the sum of Nobel laureates for the two countries is remarkably constant, equal to five for all five-year intervals between 1957 and 1982. In the first interval there were four Russians and one Swede, but this ratio reversed in a continuous way, and twenty-five years later there were four Swedes to one Russian. The evolution of the Swedish-Russian ratio on the logarithmic scale turned out to be quite compatible with a straight line, the hallmark of a natural substitution (Appendix C, Figure 6.7). Was this an artifact of statistical fluctuations? Could it be attributed to a subconscious bias, considering that the Nobel Foundation is Swedish and that traditionally there has been some animosity between the two countries? Or could it indeed be a natural substitution in a local microniche?

Two arguments come to mind, and both stand against the possibility of a microniche. The first is based on relating competitiveness to age. The average age of the Swedish laureates is 65.1, distinctly higher than the average Russian age of 57.5 during the same period. Thus, there can be no competitive advantage of a Darwinian nature relating to age for the Swedes during this period, as can perhaps exist for Americans over Europeans and, later on, for the rest of the world over the United States.

The second argument fails with my inability to justify a microniche occupied by Russians and Swedes over this particular period of time, in which a natural substitution might occur.

As a third hypothesis, it is conceivable that this phenomenon, of which we may have seen the end, considering that there have been no Swedish laureates during the last five years, could have been triggered by the unpopular Russian intervention in Hungary. Russian popularity dropped worldwide at that time and suffered further losses with the actions taken later, against Czechoslovakia in 1968 and against Afghanistan in 1978. This hypothesis is reinforced by the frankly political nature of the Nobel prize for peace, in particular, over the last several years.

FISHING AND PRYING

When Fisher and Pry put forth their model for competitive substitutions, they were mainly concerned with the diffusion of new technologies. Similar applications of the logistic function have been employed in the past by epidemiologists in order to describe the spreading of epidemic diseases. It is evident that the rate of new victims during an epidemic outbreak is proportional to both the number of people infected and the number remaining healthy, which spells out the same law as the one describing natural growth in competition. The diffusion of an epidemic is a substitution process in which a healthy population is progressively replaced by a sick one. In all three processes, diffusion, substitution, and competitive growth, the entities under consideration obey the same law.

In their paper Fisher and Pry looked at the rate of substitution for a variety of applications and found that the speed at which a substitution takes place is not simply related to the improvements in technology or manufacturing or marketing or distribution or any other single factor. It is rather a measure of how much the new is better than the old in *all* these factors. When a substitution begins, the new product, process, or service struggles hard to improve and demonstrate its advantages over the old one. As the newcomer finds recognition by achieving a small percentage of the market, the threatened element redoubles its efforts to maintain or improve its position. Thus, the pace of innovation may increase significantly during the course of a substitution struggle. The curvature of the bends and the steepness of the S-curve traced out by the ratio of total market shares, however—the slope of the straight line in the logarithmic graph—does not change throughout the substitution. The rate reflected in this slope seems to be determined by the combination of economic forces stemming from the superiority of the newcomer.

Fisher and Pry suggested that this model could prove useful to investigations in the many aspects of technological change and innovation in our society. In fact, its use has spread far beyond that application. I do not know if they realized back in 1970 the diversity of the areas of inquiry into which their model would diffuse twenty years later, but I agree with Marchetti when he says that they provided us with a tool for "Fishing and Prying" into the mechanisms of social living.

7

Competition Is the Creator and the Regulator

The original saying is attributed to the Greek philosopher Heraclitus. Translated from the ancient Greek it says: "War is the father and the king of everything." However, the meaning of war goes well beyond the customary image of violent conflict. Heraclitus, who has been characterized as the first Western thinker, sees war as a divine force, a natural law that distinguishes gods and people, declares masters and slaves;[1] a law that creates and controls what happens. In that sense such a law is more appropriately interpreted as competition than ordinary warfare.

Twenty-five hundred years later Charles Darwin arrived at conclusions of comparable importance concerning competition when he formulated the principle of survival of the fittest, otherwise known as natural selection. In so doing Darwin provoked the wrath of contemporary churchmen but the joy of biologists who built a successful science on his ideas. Mathematicians such as Vito Volterra and Alfred J. Lotka put this theory into equations capable of describing the intricate interaction between predator and prey in species populations. At the heart of this formulation lies the *logistic function,* the mathematical representation of a population growing under competition.

Natural selection did well in describing the growth of species populations. Sociologists and psychologists did not take long to start borrowing this idea in order to model human competitive behavior along similar lines. Eventually the competition formalism was brought to inanimate, more abstract domains, such as products, industries, primary energies, means of transportation, and diseases. The presence of competition in these domains had long been recognized but not with the mathematics capable of making quantitative projections which would provide new insights and produce conclusions that are often interesting and sometimes far-reaching.

Evolution of a species through natural selection is possible thanks to mutations and the right conditions. Mutations take place all the time, but under normal circumstances they are eliminated because they are ill fit to survive. The species has already reached an optimal configuration through a long period of natural selection; random mutations are unlikely to produce a competitive advantage, so they perish. The continuous availability of mutations, however, serves an important role at times of violent environmental change to which the species has to adapt or become extinct. The pool of mutants is then used to select those with characteristics best fit for survival. The larger the pool, the bigger the chance to adapt successfully. The species undergoes rapid evolutionary changes until a new optimization is achieved and the species' strategy becomes conservation once again.

A similar situation is encountered in industry. A company or a whole industry that has been optimized becomes conservative. You do not want to change something that works well. However, a portfolio of mutations is kept in store and will come out only in times of crisis. Innovation, reorganizations, acquisitions, and other sometimes erratic actions are all geared toward reoptimization. The organization hunts for the optimal configuration to ensure its survival.

The steel industry, for example, has been conservative for a long time, but now it may finally have reached the crisis situation, facing stiff competition from alternative new materials. In contrast, the electronics industry is still very mutational. It behaves as a young industry searching its way to optimization, which will ensure better chances for survival. It also tries to increase the number of directions it explores, for overspecialization may endanger survival. Pandas have become an endangered species because they eat only bamboo, which is in scarce supply. Sharks are better optimized; they can eat just about anything. Environmentalists do not worry about sharks, at least not for the same reasons.

Long-term survival involves alternating between conservative and innovative behavior. If things go well, a winning strategy is "change nothing." Following radical changes in the environment, however, the appropriate policy becomes "look around for new directions, explore many different ideas, in order to increase your chances of falling on a good track again."

As we have seen in previous chapters, natural growth in competition can be described "logistically" on an S-curve. Richard Foster, a director of McKinsey & Company, a leading management consulting firm, promises success in business if managers promptly adapt their strategy according to where their companies are placed on the S-curve of the particular growth process they happen to be going through. He urges company executives to switch resources ruthlessly when growth reaches saturation and the time is ripe for innovation.[2] Marc van der Erve from Digital Equipment Corporation develops this further by defining the role of a "culture performance driver," someone whose job it is to influence the inclination of people within a corporation to be conservative or to abandon the old ways and search for new ones. He believes this can be done through two types of "culture forces." One force encourages "jumping S-curves," and the other one motivates people to hang on to the established ones. The timely application of these forces ensures success. Too early, you may miss out on profits; too late, you may go bankrupt before implementing the change.[3]

A succession of growth phases can be visualized as cascading S-curves. Each phase is a small-scale natural-growth process itself, and it may well be that a lifelong overall S-curve is composed of many small ones. A new growth process begins before the old one ends and continues growing after the old one reaches its ceiling. The succession from one S-curve to the next is shown schematically in the upper portion of Figure 7.1. The steeply rising section is the time of rapid and successful growth, and therefore a time of conservatism. The flattening of the curve indicates a decline in the rate of growth, and when it becomes significant, it is time for change. The lower portion of Figure 7.1 shows the life cycles—the rates of growth—that correspond to the cascading growth processes.

It should not come as a surprise that the seeds of a new cycle are sown right after the peak of the previous one. The emerging pattern, a succession of wave crests, suggests that sustained growth and evolution itself is not a flat uniform process.

SUCCESSION OF GROWTH PROCESSES

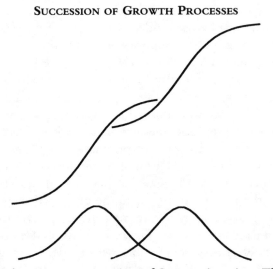

FIGURE 7.1   Schematic representation of S-curve jumping. The bell-shaped curves at the bottom are the life cycles that correspond to the processes.

WHEN IT RAINS, IT POURS

A succession of life cycles, as indicated at the bottom of Figure 7.1, traces a wavy outline. With life cycles representing the rate of growth (for example, the number of units of a particular product sold per month), the peaks of the waves imply that there is a greater frequency of occurrences at that time. A many-crested wave implies successive periods of clusters of occurrences. Proverbs in many cultures have paid tribute to the phenomenon of clustering, which may be contested by statistician purists. Events that normally should be randomly distributed over time appear in well-defined clusters followed by periods when they do not appear at all. The superstitious will tell you that bad things always occur in threes or "it never rains but it pours." For some, good (or bad) luck seems to come in waves when they are either "on a roll" or "under a black cloud."

Peaks of luck at casino roulette wheels can easily be demystified by mathematicians invoking the laws of statistics and the possibility of rigged roulette wheels. Peaks of success with the opposite sex are explained by psychologists as a high in a person's general well-being that results in increased sex appeal, which attracts many of the opposite sex in a short time. What is perhaps less easily explained is the clustering that has been observed in the history of discoveries, in innovation, in riots and political unrest, and in violence and warfare.

The first time I was impressed by such clustering of otherwise random events was when I read a book titled *Stalemate in Technology* by Gerhard Mensch in which there was a graph showing the time and number of the appearances of all basic innovations in Western Society. Mensch classified what he regarded as basic innovations over the last two hundred years and found that they did not come into existence at a steady rate but at a rate that went through well-distinguished periods of peaks and valleys.

Mensch defines as a basic innovation something that starts a new industry or results in a new kind of product—the phonograph, for example. Each innovation is based on an invention or discovery that has been in existence for a while and that makes it all possible. Later improvements in the manufacturing or in the quality of these products are not counted as innovations. A clustering pattern emerges from Mensch's classification of basic innovations. Figure 7.2 is adapted from his book and shows the number of basic innovations per decade. Four clear peaks can be distinguished, spaced rather regularly fifty to sixty years apart.

There is a certain amount of arbitrariness in Mensch's definition of

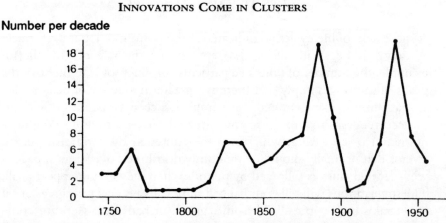

INNOVATIONS COME IN CLUSTERS

Number per decade

FIGURE 7.2   The data points represent the number of basic innovations per decade according to Gerhard Mensch's classification. A dramatic variation of the number of innovations over time becomes evident even if the exact number of innovations may be subject to debate.*

*Adapted from a graph by Gerhard Mensch in *Stalemate in Technology: Innovations Overcome the Depression* (Cambridge, MA: Ballinger, 1979). Reprinted by permission of the publisher. See also the German edition, *Das technologische Patt* (Frankfurt: Umschau Verlag, 1975). The German edition is preferable for details on how the data were selected.

basic innovations. The emerging clustering pattern, however, persists among several other attempts at such classification.[4] It has also been noted that innovations appear to be seasonal, like agricultural crops. Throughout the winter after harvest a fruit tree undergoes a slow nurturing process, and in the spring its branches blossom in order to produce the next crop. According to Marchetti, innovations in a social system are analogous: The social system plays the role of the tree, and the innovations that of its fruit.[5]

Even though Mensch's conclusion was corroborated by other classifications of basic innovations, I still had reservations because of the subjective element ingrained in procedures such as tabulating quantitatively two centuries' worth of innovations. Yet the idea of clustering was appealing to me, and its validity was reinforced when I came across a less subjective case in which there was no room for ambiguity with regard to definitions, units, or dates: the discovery of the stable chemical elements, a subject with which I was intimately involved throughout my career as a physicist. Figure 7.3 depicts the discovery of the stable elements over the centuries. The graph at the top shows the number of elements known at any given time. In Chapter Two it was suggested that a single S-curve can approximately describe the overall set of data points in the discovery of stable elements. Taking a closer look here we may distinguish four smaller S-curves in cascade. The lower graph shows the life cycles corresponding to each period, depicting the number of elements discovered per decade. The emerging picture presents peaks of activity similar to those seen in Mensch's classification of basic innovations.[6]

It is not difficult to come up with an explanation for this clustering. The first twelve elements have been known from antiquity: gold, silver, iron, copper, carbon, and so forth. Their discovery may have followed an initial growth process stretching out over many centuries. Discovering the remaining elements began with industrialization in the middle of the eighteenth century and progressed in well-defined cycles. Each cycle is associated with the technology available for separating these elements. The cycle of using chemical properties was succeeded by the cycles of using physical properties. This was succeeded by the cycle of using nuclear properties, whereupon elements started being created artificially in particle accelerators, which produced elements that decay soon after creation and hardly deserve the name stable. At any rate this cycle also seems practically completed; forecasts about the number of elements yet to be discovered or created must be limited to one or two. The succession of cycles in element discovery must be considered at an end.

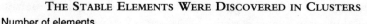

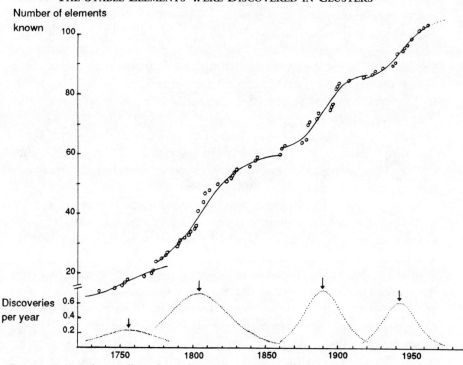

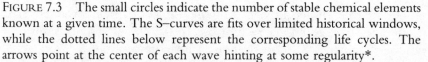

FIGURE 7.3   The small circles indicate the number of stable chemical elements known at a given time. The S–curves are fits over limited historical windows, while the dotted lines below represent the corresponding life cycles. The arrows point at the center of each wave hinting at some regularity*.

A question remains, however. Why are the cycles almost regularly spaced? Is there any significance to the fact that this spacing is similar to the one that characterized the appearance of basic innovations? We will attempt to answer this question in the next chapter. For the time being, Figure 7.3 provides evidence that an overall S-growth may reveal, upon closer examination, a succession of similar but shorter contours reflecting a wavy structure rather than a uniform continuous stream.

S-curve succession was first mentioned in Chapter Two with the example of an infant's vocabulary, which reaches the first plateau at the age of six when it exhausts the vocabulary of the home "niche." Acquisition of vocabulary (like the discovery of stable elements) proceeds in waves, one before six, and one or more afterward depending on the

* The data came from The American Institute of Physics Handbook, 3rd ed. (New York: McGraw-Hill).

linguistic variety of the subsequent environments. The home niche is followed by the school niche, which may be followed by yet another niche later on.

But one may need to combine S-curves in a different way than by simply cascading them. There are situations where a new niche may appear within an older niche, usually after some technological change, such as making watches waterproof to allow swimmers to wear them. The situation of a niche-within-a-niche is exemplified by Alfred Hitchcock's life's work.[7] In Figure 7.4 we see a graph of the number of films for which Hitchcock could claim credit at a particular time during his career.

As a child Hitchcock manifested interest in theatrical plays, but as a teenager he started going to movies more often and soon began frequenting movie studios. At the age of twenty he took a modest job as a designer of titles for the silent movies of the time, pretending—as he

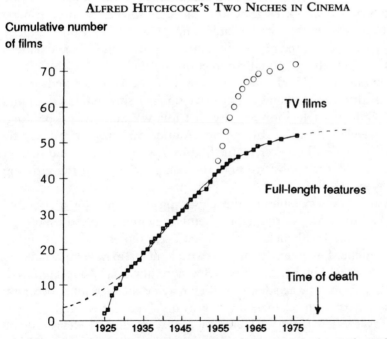

ALFRED HITCHCOCK'S TWO NICHES IN CINEMA

FIGURE 7.4 The squares indicate full-length films while the circles indicate the sum of both full-length and shorter television films. The fit is only to the full-length films. A smaller curve is outlined by the television works and seems to have its beginning in Hitchcock's film works. The configuration of these two S-curves provides a visual representation of the niche-within-a-niche situation.

maintained even later—that he had no ambition to assume more responsibility. This is in contradiction to his insistence on learning everything
there was to learn about filmmaking and volunteering to try out his hand
at any new assignment. In fact, the lower part of the curve fitted to the
data of his full-length films seems to originate well before 1925 when his
first movie appeared. When he finally started his career as a film director
at twenty-six, he produced prodigiously during the first six years, as if he
were trying to "catch up," not unlike others discussed in Chapter Four
whose early careers display this sudden release of pent-up productivity.

From 1930 onward the cumulative number of Hitchcock's full-length
features grows smoothly to reach fifty-two by 1975, 96 percent of his
potential specified by the curve's ceiling at fifty-four. Here is another case
in which little significant productive potential remained to be realized at
the time of death. The twist in Hitchcock's case is that in 1955 he was
persuaded to make films for the celebrated television series *Alfred Hitchcock Presents*. The open circles on the graph represent the sum of both the
full-length and the shorter television films. A smaller S-curve can be
clearly outlined on top of the large one. This niche-within-a niche
contains twenty films; the process of filling it up starts in 1955 and flattens
out when approaching natural completion in 1962.

The evolution of Hitchcock's work just before he embarked on the
television adventure contains a suggestive signal, a slowing down leading
smoothly into the television activity that follows. Statistically speaking,
the small deviation of the data points around 1951 has no real significance. It coincides, however, with the times when the film industry in
the United States felt most strongly the competition from the growing
popularity of television.

Growth in cycles results in the clustering phenomenon mentioned
earlier. The discovery of the stable chemical elements goes through cycles, and so does Hitchcock's work. Each cycle represents the filling up
of a new niche. Discovering one or two elements may seem a random
event at first, as Hitchcock's first television film may have appeared.
Isolated events may be accidents, or they may be the beginning of a new
wave. If something fundamentally new is involved, for example, a new
technology for separating elements or a new medium (television) for
filmmaking, one can surmise that the isolated events are not accidents but
rather the opening up of a new niche. If this niche is to be filled naturally,
then the process will go through the successive stages of growth, maturity, and decline. Consequently, more such events will follow, giving rise
to a cluster that becomes the life cycle of the process.

Clustering is observed in every growth process. If the overall time period is short, for example the diffusion of fashion or fads, the wave is felt strongly. If the process proceeds slowly, the clustering may pass unnoticed. For instance, few people are aware of the fact that life expectancy went through cycles of growth, as we will see in the next chapter: one that peaked around the turn of the century and another that peaked in the 1950s.

SUCCESSIVE SUBSTITUTIONS

It was demonstrated in Chapter Six that natural substitutions proceed along the same S-shaped patterns as the populations of species. The introduction of a gifted new competitor in an already occupied niche results in a progressive displacement of the older tenant, and the dominant role eventually passes from the old to the new. As the new gets older, it cedes leadership in its turn to a more recent contender, and substitutions thus cascade. For example, steamships replaced sailing ships, but later they themselves were replaced by ships with internal-combustion engines. While the total registered tonnage of the merchant fleet in the United States increased by a factor of almost one hundred during the last two hundred years, the percentages of the different types of ship show two successive substitutions.

In Figure 7.5 the vertical scale is again such that S-curves become straight lines. We can distinguish two one-to-one substitutions: steam for sail before 1900 and motor for steam after 1950. Between these dates all three types of boats coexist, with steam claiming the lion's share. It is worth noting that even today motor ships have not yet acquired more than 25 percent of the total tonnage. The number of steamships may be decreasing in favor of motor ships, but they still remain the dominant type of merchant vessel today; they are often fueled by oil instead of coal, and in some cases they use steam turbines.

Internal-combustion engines, when first introduced in ships, started spreading "abnormally" rapidly, probably due to the momentum acquired in their swift propagation through transportation overland (replacement of horses by automobiles), which had just preceded. Suddenly the realities of World War II, notably a shortage of gasoline, interrupted the accelerated introduction of motors in ships, and the substitution process was readjusted to a level and a rhythm that later proved to be the natural trend.

FROM SAILS TO STEAM TO MOTORS

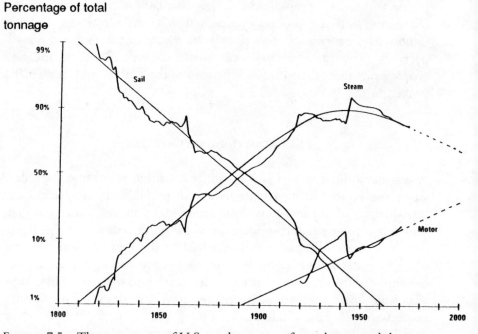

FIGURE 7.5   The percentage of U.S. total tonnage of merchant vessels by type of ship. Again, the vertical axis shows the logistic scale that transforms S-curves to straight lines. The straight lines shown are fits to the data. The curved section is the transition period between the end of phasing in and the beginning of phasing out (see Appendix A). Steamships substituted for sailboats between 1800 and 1920, at which time boats with internal-combustion engines started replacing steamships. World War II seems to have brought this substitution process down to its natural course.*

The smooth lines in Figure 7.5 represent the description given by the substitution model as generalized by Nebojsa Nakicenovic to include more than two competitors.[8] It says that at any given time the shares of all competitors *but one* will follow straight-line trajectories. The singled out competitor, called the *saturating* one, will have its share determined as the remainder after all other shares have been subtracted from 100 percent. The saturating competitor is normally the oldest among the ones that were still growing. Usually it is also the one with the largest share,

*Adapted from a graph by Nebojsa Nakicenovic in "The Automobile Road to Technological Change: Diffusion of the Automobile as a Process of Technological Substitution," *Technological Forecasting and Social Change,* vol. 29 (1986): 309–40. Copyright 1986 by Elsevier Science Publishing Co., Inc. Reprinted by permission of the publisher.

has practically reached maximum penetration, and is expected to start declining soon. The trajectory of a saturating share is curved. It traces the transition from the end of the growing phase, where the entity was substituting for its predecessor, to the beginning of the declining phase, where it will itself start being replaced by the one next in line. Every competitor goes through the saturating phase in chronological order.

The saturation phase can be seen as the culmination and, at the same time, as the beginning of the end. For a product it often corresponds to the maturity phase of the life cycle. It is the time when you face competition directly. You have done well, and everyone is trying to chip away at your gains. Your share is calculated as what remains after subtracting from 100 percent the trajectories of all the others. You are at the top, but you are alone; the model requires *one and only one* in the saturating phase at a time. This condition is necessary in order to produce a workable model, but it also matches well what happens in a multicompetitor arena—typically, the Olympic Games.

ONLY ONE FRONT-RUNNER AT A TIME

Taking care of logistics for the Olympic Games must be a feat comparable, if not superior, to the records broken during the games themselves. In anticipation of the games, preparations may include construction of housing, roads, subway lines, stadiums, and whole villages. The number of visitors to be accommodated sometimes surpasses the number of local residents.

Far easier to decide is the number of medals to be fabricated. This proof of distinction comes in three versions: gold, silver, and bronze. And there must be exactly one of each for each event in the games, no more, no less. Among the world's most exceptional athletes there can be no surprises about the enumeration of the winning performances. For each event there will be a first, a second, and a third. In every event the three winners outperform the average mortal by the same factor. Nevertheless, under competition, one can distinguish among them as needed by using precision watches with time resolutions of $\frac{1}{100}$ of a second or better. The difference between first-, second-, and third-place medals seems exaggerated compared to the difference between performances.

• • •

In a typical American living room the family has gathered around the television, once more watching the Olympic Games.

"Look, Mom, they're going to cry again," says the little one.

It is the show's most titillating moment, awards time. Three young athletes step on the pedestals of different height. Gold, silver, and bronze medals are hung around the appropriate necks. Emotions are high both in the stadium and in the living room. All three athletes are crying.

"They are crying from joy," says Mom.

"The second and third don't seem joyful," objects the little one.

• • •

It may well be the case. Second and third places are usually not coveted by competitive individuals. Everyone competes for the first position. The nature of competition is such that there is always *one* front-runner, and everyone runs against him or her.

Charting this phenomenon, the generalized substitution model with its logarithmic scale gives rise to the image of a mountainous landscape similar to Figure 7.5. Crisscrossing straight lines denote natural substitutions, and a curved peak designates the front-runner of the period, someone who was aspiring to be first only a short time ago and who will not stay first for long.

Detailed below are two examples of this process: means of transportation and primary energy sources. They use yearly data from the *Historical Statistics of the United States*[9] spanning the last two centuries.

TRANSPORT INFRASTRUCTURES

Means of transportation have evolved toward an ever higher performance. Increasing speed is one obvious improvement, but significant jumps in performance must be seen in terms of productivity (speed times payload) measured in terms of ton-mile per hour or passenger-mile per hour.

The first major improvement in the U.S. transport system was the construction of canals aimed at connecting coastal and inland waterways in one infrastructure. Canal construction started about two hundred years ago and lasted almost one hundred years. By the end of the nineteenth century some links were already being decommissioned because traffic began moving to railroads. The first railways were constructed in 1830, increasing speed and productivity. The railway's prime lasted until the

1920s when trains started losing market share to automotive transport. The total length of railway track *in use* has actually decreased since then by 30 percent. The automobile came into wider use around the turn of the century and, coupled with the availability of paved roads, increased speed and performance once again. Finally, air travel entered the scene in the 1930s, and airway connections between cities began to be established. Despite the fact that air routes can be opened and closed more easily than railway and highway connections, the mileage of the different transport systems can be meaningfully compared because they all serve the same purpose.

Each successive transport infrastructure—canals, railways, paved roads, and airways—provided about one order of magnitude improvement in the average productivity. In that light a future means of transportation cannot be supersonic transport as it is known today even if it achieved a speed of several Mach (one Mach equals the speed of sound). These airplanes may achieve higher speeds, but in fact they decrease the productive payload. The success of the European Concorde was compromised by the fact that the time it gained flying, it lost refueling.

Aviation know-how has already achieved maturity today, and simple technological advances cannot produce factors of ten improvements. A new technology, fundamentally different, is required—for example, airplanes fueled with liquid hydrogen or magnetic levitation trains.[10] The new competitor must possess an indisputable competitive advantage. The scrapping of the supersonic transport project (SST) back in 1971 by the Senate displeased the Nixon administration but may have been symptomatic of fundamental, if unconscious, reasoning. The comment by then Senator Henry M. Jackson, "This is a vote against science and technology," can be seen today as simplistic and insensitive to rising popular wisdom.

But let us get back to the successive substitution between the different transport infrastructures. A traveler or a package can make a trip partly by road, partly by rail, and partly by air. The various means of transportation are functionally interconnected and can be seen as forming an overall network. The growth of the total length of this network over the past 180 years has followed an S-curve and is today at 80 percent of its final size, as Grubler shows.[11] Each separate infrastructure can be expressed as a fraction of this total. Each system enters, grows, reaches a phase of maturity when it commands a dominant share (more than 80 percent) of the total "market," and then declines. Curiously, the phase of maturity is reached long before the end of construction. The total railway track grew by a factor of ten between the time railways enjoyed maximum share in the competitive picture and the time construction finally ceased.

In the generalized substitution model the succession of the different transportation infrastructures outlines a typical mountainous landscape (Appendix C, Figure 7.1). From the beginning of the nineteenth century the share of canals declined along a straight line in favor of railways, whose share reached a peak around 1880, commanding more than 80 percent of all transport mileage, with the remainder split equally between canals and roads. As the road mileage increased rapidly the share of railways declined, despite continuous growth in the length of railway track. The early twentieth century witnessed a one-to-one substitution between transportation by rail and transportation by road. The growth in relative importance of paved roads reached saturation by the 1960s when its share reached more than 80 percent, and railways split the remainder with airways. Finally, airways "phased in," claiming a significant share by the second half of the twentieth century, forcing paved roads to enter a relative decline.

In the case of air transport the "length" is defined as the total route mileage operated by the airways. This estimate, based on an analogy to the *physical* lengths of the other infrastructures, is rather crude. Air routes are opened and closed in response to market demand, while such a response usually represented a major undertaking for the builders of canals, railway tracks, and paved roads. Still, the fact that the data agree well with the substitution-model description comes as a confirmation of our hypotheses.

The successive declines observed are not due to the physical destruction of the infrastructure's mileage. "Phasing out" is relative and due mainly to the fact that each new means of transportation increased the overall mileage length by one to two orders of magnitude. Most canals are still around today, but other uses have become fashionable, such as leisure activities, transport of low-value goods, and irrigation. The situation is similar in the case of sea transport. Nakicenovic points out that there are more sails fabricated for sailboats today than in the days of ocean clippers, but most of them are used for pleasure and they do not contribute to the transportation of a single commercial ton-mile.

THE PRIMARY ENERGY PICTURE

Another market where various technologies coexist and compete with each other is that of primary energy sources. This is Marchetti's oldest, favorite and most often cited example, first published in 1975. He wrote then: "I started from the somehow iconoclastic hypothesis that the different primary energy sources are commodities competing for a market, like different brands of soap . . . so that the rules of the game may after all be the same."[12]

Marchetti then applied the generalized substitution model and found a surprisingly good description of worldwide data over a historical window of more than one hundred years. His updated results are reproduced in Figure 7.6. During the last one hundred years, wood, coal, natural gas, and nuclear energy are the main protagonists in supplying the world with energy. More than one energy source is present at any time, but the leading role passes from one to the other. Wind power and water power

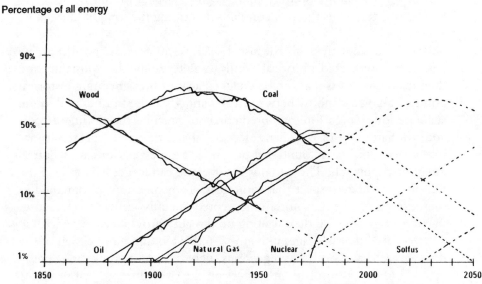

FIGURE 7.6   Data, fits, and projections for the shares of different primary energy sources consumed worldwide. For nuclear, the dotted straight line is not a fit but a trajectory suggested by analogy. The futuristic source labeled "solfus" may involve solar energy and thermonuclear fusion.*

*Adapted from a graph by Cesare Marchetti in "Infrastructures for Movement," *Technological Forecasting and Social Change,* vol. 32, no. 4 (1987):373–93. Copyright 1987 by Elsevier Science Publishing. Reprinted by permission of the publisher. The graph originally appeared in Nebojsa Nakicenovic, "Growth to Limits," (Ph.D. diss., University of Vienna, 1984).

provide an amount of energy that represents less than 1 percent of the total and are not visible in the figure.

In the early nineteenth century and before, most of the world's energy needs were satisfied through wood burning and to a lesser extent animal power not shown in the figure. Contrary to the popular image of coal-burning locomotives, wood remained the principal fuel for railroads in the United States up to the 1870s. The substitution process shows that the major energy source between 1870 and 1950 was coal. Oil became the dominant player from 1940 onward, as the automobile matured, together with petrochemical and other oil-based industries.

It becomes evident from this picture that a century-long history of an energy source can be described quite well—smooth lines—with only two constants, those required to define a straight line. (The curved sections are calculated by subtracting the straight lines from 100 percent.) The destiny of an energy source, then, seems to be cast during its early childhood, as soon as the two constants describing the straight line can be determined.

Marchetti goes through the usual exercise of forecasters; that is, he considers a restricted historical window, determines the constants, and then makes forecasts that can be checked against the remaining historical points. Taking a window between 1900 and 1920, he finishes with quite accurate long-range forecasts predicting successfully the decline of coal and the flattening of oil, many decades ahead of time. The results are impressive, but I have found that pragmatic business persons are skeptical about this approach. They do not believe one can forget having seen the full picture. They suspect that some bias, if unconscious, is ingrained in such exercises. The only criterion accepted unequivocally is track record, making a forecast, and then waiting until reality proves it right or wrong.

There are other messages in Figure 7.6. By looking more closely at the data we see that world-shaking events such as wars, skyrocketing energy prices, and depression had little effect on the overall trends. More visible are the effect of strikes. In the coal industry, for example, such actions result in short-term deviations, but the previous trend is quickly resumed.

Another observation is that there is no relationship between the utilization and the reserves of a primary energy source. It seems that the market moves away from a certain primary energy source long before it becomes exhausted, at least at the world level. And vice versa; despite the ominous predictions made in the 1950s that oil would dry up in twenty years, oil use continued growing unhindered and more oil was found as needed. Oil reserves will probably never be exhausted because of the

timely introduction of other energy sources. Well-established substitution processes with longtime constants are of a fundamental nature and
cannot be reversed by "lesser" reasons such as depletion of reserves.[13]

Figure 7.6 also indicates that natural gas will replace oil progressively to
reach a zenith in the 2020s and become more important worldwide than
oil was in the 1970s. Supplying a major fraction of the world's energy
needs by gas will require much more gas than today's proven reserves, but
one need not worry about it; important natural gas fields are likely to be
found. Searches for "dry" gas have started relatively recently. Gas is a
more probable find than oil the deeper one goes underground, due to the
thermal gradient of the earth's crust. Ultimately, if during the gas era the
discovery of new gas fields does not keep up with demand, for whatever
reason, oil or coal may be artificially processed to produce the amount of
gas lacking. Synthetic gaseous fuels such as methanol could easily be used
in cars during the twenty-first century.

The forecast of the takeover by natural gas was first made in 1974 at a
time when experts disagreed with such a notion.[14] Half a generation later
the environmentalists have converged on the conclusion that natural gas
is "the miracle energy source for the future." In 1990, George H. Lawrence, president of the American Gas Association, was quoted as predicting that gas consumption nationwide would "rise by one-third over
the next twenty years."[15] I do not know what methodology he used for
his estimate, but if I were to use the energy-substitution picture and the
fact that world energy consumption has been growing at a steady average
of 2 percent annually since 1860, I would find a 220 percent increase in
natural gas consumption twenty years from now. In the United States
alone the increase is certainly bigger because energy consumption here
has been growing at least 3 percent annually. Doing the analysis quantitatively for the United States I find that by the year 2000 natural gas
consumption should be four times that of the year 1982. For this to
happen we must soon witness an impressive rush toward the use of
natural gas.

As the role of natural gas acquires importance, governments and large
corporations will be engaging in power struggles over it. Ken Weiner,
Jimmy Carter's deputy director of the Council for Environmental Quality and now a Seattle attorney, remarked recently: "It has been said war
is too important to be left to the generals. Some are wondering if environment quality is too important to be left to the environmentalists."[16] In
light of that remark it is perhaps worth raising the question of what has
really been the role of environmentalists in the greening of natural gas.

The importance of gas in the world market has been growing steadily for the last ninety years, independent of latter-day environmental concerns. The voice of environmentalists reminds me of Ralph Nader's crusade in the 1960s for car safety, while the number of deaths from car accidents had already been pinned between twenty-two and twenty-eight annually per one hundred thousand population for more than forty years (Figure 1.1).

Environmentalists have also taken a vehement stand on the issue of nuclear energy. This primary energy source entered the world market in the mid 1970s when it reached more than a 1 percent share. The rate of growth during the first decade seems disproportionally rapid, however, compared to the entry and exit slopes of wood, coal, oil, and natural gas, all of which conform closely to a more gradual rate (see Figure 7.6). At the same time the opposition to nuclear energy also seems out of proportion when compared to that of other environmental issues. As a consequence of the intense criticism, nuclear energy growth has slowed considerably, and I would not be surprised to find the data points of the next few decades more along the straight line proposed by the model in the figure. One may ask what was the prime mover here—the environmental concerns that succeeded in slowing the rate of growth or the nuclear energy craze that forced environmentalists to react?

The coming to life of such a craze is understandable. Nuclear energy made a world-shaking appearance in the closing act of World War II by demonstrating the human ability to access superhuman powers. I use the word superhuman because nuclear reactions are based on mechanisms through which stars generate their energy. Humans for the first time possessed the sources of power that feed our sun, which was often considered a god in the past. At the same time mankind acquired independence; nuclear is the only energy source that would remain indefinitely at our disposal if the sun went out.

Figure 7.6 suggests that nuclear energy has a long future. Its share should grow at a slower rate, with a trajectory parallel to those of oil, coal, and natural gas. But there is no alternative in sight. The next primary energy source—fusion and/or solar and/or other—is projected to enter the picture by supplying 1 percent of the world's needs in the 2020s. This projection is reasonable because such a technology, once shown to be feasible, would require about a generation to be mastered industrially, as was the case with nuclear energy.

Let us try to be excessively optimistic and suppose that technology gallops during the next ten years and a new clean energy source enters the world market as early as 2000. It will have to grow at the normal rate, the rate at which other types of energy have entered and exited in the past. Therefore, it will have an impact on nuclear energy similar to the way gas had an impact on oil. That is, nuclear energy, and to a lesser extent gas, will saturate at lower levels while the new energy will reach relatively higher shares. Still, nuclear energy will have played a role at least as important as oil did in its time.

But another reason I believe this optimistic scenario is unrealistic, apart from matters of technology, is that it would make the gas and nuclear peaks come earlier and consequently fall out of step with the fifty-six year cycle that seems to dominate such cosmic matters. We will be looking at this cosmic cycle closely in the next chapter, in particular in Figure 8.3.

THE DISEASE "MARKET"

Here is yet another example of successive substitutions along natural-growth curves. I mentioned earlier that diseases can be seen as competing for the biggest share of all deaths. Young diseases are on the rise while old ones are phasing out. Charting this phenomenon I obtained a simplified but quantitative picture in which diseases were grouped into three broad categories according to prevalence and nature (Appendix C, Figure 7.2). One group contained all cardiovascular ailments, today's number-one killer. A second group was called neoplastic and included all types of cancer. The third group comprised all old, generally phasing out, diseases. Hepatitis and diabetes were not included because their share was rather flat over time (no competitive substitutions) and below the 1 percent level (not visible in the usual picture). Cirrhosis of the liver also represented too small a percentage, but being of an intriguing cyclical nature, it is reported separately later in Figure 8.4.

The disease "market" presented the classic mountainous landscape of the dynamics of natural competition. From the turn of the century, cardiovascular ailments have been claiming a progressively increasing share, making them the dominant cause of death by 1925. Their share reached a peak in the 1960s with more than 70 percent of all deaths and then started declining in favor of cancer, which had also been steadily growing at a comparable rate—parallel line—but at a lower level. All other causes of death had a declining share from the beginning of the century. Today cardiovascular ailments may claim twice as many victims as cancer, but their share is steadily decreasing in favor of the latter.

Projections of the natural trajectories indicate that cancer and cardiovascular diseases will split the total death toll fifty-fifty by the year 2010.

It is likely that in the future new diseases will enter the scene. The most frequently cited candidate today is AIDS. In my picture I had no trend yet for AIDS—the historical window being too short—so I created an exaggerated scenario in which AIDS grew toward filling the whole niche. In 1988 AIDS was at the level of 0.7 percent of all deaths. Beginning with this data point I imposed on the growth of AIDS a slope comparable to those of cancer and cardiovascular ailments, and that placed the "official" entry of AIDS—the 1 percent level—around the year 2000. The growing new competitor caused the share of cancer to saturate, go over a peak during the second half of the twenty-first century, and enter decline by the twenty-second century, as AIDS attained today's levels of cancer.

The above projection for the evolution of AIDS makes its threat much more important than the conclusion drawn in Chapter Five, namely that AIDS is confined to a microniche of 1 percent of all deaths. Furthermore, the assumption that it will grow at the same rate as cardiovascular ailments did in the beginning of the century may be challenged on the grounds that AIDS is an *epidemic* disease with different mechanisms of diffusion. This is true, but it is also true that the person who wants to avoid AIDS at any cost can do so fairly reliably, which is not the case with heart diseases. Under these circumstances one may want to believe that when the AIDS threat becomes sufficiently menacing, human behavior will progressively adapt itself so that the final trajectory features the "normal" slope after all, or rather that the disease is indeed confined to a small micro-niche.

The scenario that projects the development of AIDS into a major cause of death is an academic exercise suitable for simulating the onslaught of new disease. The actual data on AIDS indicate that the tolerance level acceptable to American society for this enemy may be closer to the 1 percent of all deaths evidenced in Chapter Five. Barring the appearance of yet unknown new diseases, the best forecast is that the share of cancer will continue to grow unhindered.

NOBEL PRIZE AWARDS

The phenomenon of natural substitution can also be seen in operation with a more noble form of competition: Nobel prize awards. The total number of Nobel awards per year can be thought of as a market, a pie to be distributed among the candidates. If we group together individuals with common characteristics or affiliations, we obtain regions or countries that become the contenders competing for prizes. At the end of 1987, for example, there were a total of 652 Nobel laureates from some forty countries. Many countries are represented by only a few award winners each, so grouping countries together becomes essential if one wants to look for trends. The most reasonable grouping comprises three regions. The United States is one such region. Another is "classical" Europe, consisting of Austria, Belgium, France, Germany, Great Britain, Holland, Ireland, Italy, Scandinavia, Spain, Switzerland, and the USSR. The third group is what remains and may be called the Other World. It includes many Third World countries and many developing countries, but also Japan, Australia, and Canada.

This classification must be characterized as natural because when we graph the shares of the three groups as a function of time, we see straight lines emerging (in a graph with a logistic vertical scale) for the various substitutions; and there are three such substitutions (Appendix C, Figure 7.3). The share of classical Europe accounting for 100 percent at the beginning of the century decreased along a straight line during the first four decades of the century. The United States picked up the European losses, and its share rose along a complementary straight line. After World War II the American share stopped growing, the European share shifted to a less dramatic decline, and the Other World became the rising new contender. For the last twenty years the curve of the American share has been slowly flattening out, entering by the year 2000 a decline comparable to that of the Europeans. Thus, despite the conclusion in Chapter Five that the yearly number of laureates is already decreasing for the United States, decline in competitiveness should be postponed until after the year 2000. Projecting the lines of the generalized substitution model, one finds that the Other World will overtake the Europeans in 2010 and the Americans in 2015.

The competitive substitutions described here corroborate an age advantage. The American award winners were gaining on the Europeans up to 1940, with an average age of 52.9 versus 54.8 years. Similarly, in the last ten years Other World winners are gaining on the Americans, with an average age of 57.3 versus 59.3 years, while European winners continue losing with an average of 62.4. The argument of age is more relevant when we consider that the actual work was often done many years before receiving the award. The recipient's age at award time was chosen simply because it was more readily available and because in some cases it is better defined than the date when the actual work was done. The evidence for the age advantage is correct in any case as long as there is no bias among the three regional groups with respect to this delay. Youth, once again, seems to have an advantage over old age.

Competition "reigns" in all the examples of substitution considered in this chapter. The simultaneous presence of more than two competitors in a niche gives rise to a succession of one-to-one substitutions. In each substitution the share of the old competitor declines while the share of the new competitor rises. The element of competition makes ascending and descending trajectories follow natural-growth curves and thus become predictable. From a partial set of data one can determine the lines for the trajectories of shares, which can then be extrapolated backward and forward in time. The only instance when the share of a given competitor does not follow a natural-growth curve is between the end of growth and the beginning of decline. But the shares of all other competitors do follow natural-growth curves during this time, and the sum of all shares must equal 100 percent. Therefore, even during this period, the trajectory can be calculated and, if there are no data, predicted.

8

A Cosmic Heartbeat

Energy is the ultimate food of life. More than that, one may argue that energy is responsible for the very creation of life. Consider the statistical law that says what can happen will happen if you wait long enough. Such a law acting in an energy-rich environment should ultimately create life even in the absence of consciousness. This is a philosophical thesis that deserves more development but falls outside the scope of this book. What is of interest here is the detailed way in which life uses energy and, in particular, the way humans consume energy, as well as how and why this consumption may vary over time.

Once again the United States provides the data. This country is among the youngest in history and the fastest in development, but the oldest in keeping good records of historical data. This last aspect, in view of the increasing importance that rich historical data banks have in forecasting, may prove to be a blessing.

In *Historical Statistics of the United States* we find detailed information on how energy has been spent in this country over the past two centuries. The data indicate a growth for energy consumption that is impressive. Is it exponential? Only in appearance. Natural growth is not exponential. Explosions are exponential. Natural growth follows S-curves. In their early stages, however, S-curves and exponentials appear to be very similar. Fitting an S-curve to the energy consumption data we see that it is

147

still at the beginning of its growth; the annual figures for the United
States have been following an S-curve that starts around 1850 and does
not approach a ceiling until toward the end of the twenty-first century
(Appendix C, Figure 8.1).

The fact that energy consumption may stop growing by the middle of
the twenty-second century, stabilizing at about 2.5 times today's levels,
has little impact on us today. What does have a large impact is something
that is hardly noticeable in the picture described above: the small deviations of the data points around the smooth trend. We can zoom in on
these deviations by looking at the ratio of each data point to the corresponding level of the curve. Doing so, the overall trend fades from view,
highlighting the percentage of deviation from the established pattern of
overall growth. The resulting graph, Figure 8.1, presents a picture of
regular oscillations. It is so regular that a harmonic wave—a sinusoidal—
with a fifty-six-year time period can be made to pass very closely to most
points. It indicates that while Americans consume more and more energy, sometimes they behave like gluttons toward this celestial food and

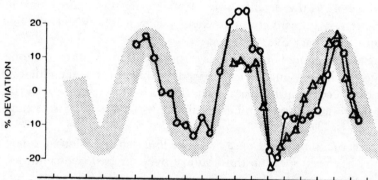

ENERGY CONSUMPTION DEPARTED FROM NATURAL GROWTH
IN A PERIODIC WAY

O   TOTAL PRIMARY ENERGY
Δ   ELECTRICAL ENERGY

FIGURE 8.1   The data points represent the percentage deviation of energy
consumption in the United States from the natural-growth trend indicated by a
fitted S-curve. The shaded "snake" is an 8% band around a regular variation
with a period of 56 years.

at other times they go on a diet. It also says that their gluttony is regularly periodic.

This periodicity in energy consumption was first observed by Hugh B. Stewart.[1] On several occasions before and after Stewart, economists and others have pointed out many human activities that oscillate within a period of fifty to sixty years.[2] It appears that it is not just Americans consuming energy who behave this way; the whole world seems to be pulsing to this rhythm. Drawing from a variety of data sources, I present in the following figure striking examples that cover widely different human endeavors. The energy consumption cycle is reproduced at the top of the figure to serve as a clock. A shaded "snake" is superimposed to guide the eye through the fifty-six-year oscillation.

The first example is the use of horsepower in the United States. Again in *Historical Statistics*[3] we can find data on the horsepower employment of all prime movers. Prime movers are defined as machines that do their work through the use of primary energy. Cars, boats, and airplanes are all good examples of prime movers, as well as electric generators and turbines in factories. Even horse-drawn carriages are prime movers. However, electric trains that use electricity produced from some other source of primary energy are not classified as prime movers.

Adding up the total horsepower defined this way, we find spectacular growth during the last century and a half. But is there any pattern in the deviations from a natural-growth curve? To find out we can make an analysis similar to the one done for energy consumption, namely, compare the data to the trend line. It again reveals a cyclical pattern that closely follows the energy consumption cycle. This may not come as a surprise. If Americans are consuming energy excessively at times, the use of horsepower will be excessive at the same times since a large fraction of primary energy goes into horsepower machines. What is less obvious is which one comes first. Does an excess in available energy result in the use of extra horsepower, or does an increase in greed for horsepower provoke an excess in energy consumption?

One may be tempted to take sides on this question. However, after looking at all the other phenomena described below, which also resonate with the same rhythm, we may consider other alternatives; for example, that neither the use of horsepower nor the consumption of energy comes first but rather that both of them respond to a common external stimulus.

As a second example we can take the clustering of basic innovations discussed in Chapter Seven. Still, in Figure 8.2 we see that the peaks in innovation coincide with the valleys in energy consumption. It may be

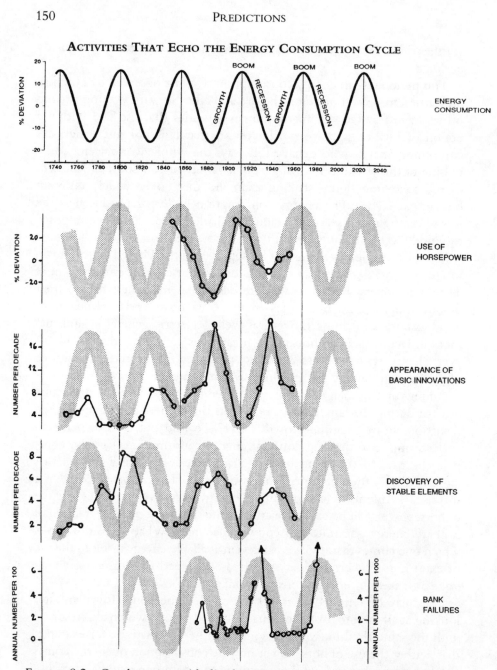

FIGURE 8.2   On the top, an idealized energy consumption cycle is used as a clock. The data concerning the use of machines represent once again the percent deviation from the fitted trend. Innovations and stable element discoveries are reported as occurrences per decade. There are two sets of data concerning banks, bank suspensions before 1933, and banks closed due to financial difficulties afterward.

out of phase, but energy consumption in America is resonating with the making of basic innovations (most of them in Europe). Are innovations linked in some way to the amount of energy consumed? They are certainly linked to discoveries, and the discovery of stable elements—the next graph shown in the figure—also seems to pulsate with roughly the same beat. The data are the same as in Figure 7.2 of the last chapter.

Here again one may be tempted to establish a causal relationship. The technologies for discovering and separating elements could themselves be linked to the basic innovations and therefore would display the same pattern. The picture becomes more intriguing, however, when one notices that innovation and discoveries peak during periods of recession and disinvestment, as can be evidenced by the graph at the bottom that charts bank failures. The data on banks for the two periods shown come from different sources. For 1933 and the few preceding years the percentage of failures is unreliable. One thing is known, however: The number was high. Fifty-six years later, circa the late 1980s, the percentage is increasing in an alarmingly similar way. For 1988 the data point is indicated with an arrow because it goes out of scale. That year 1.7 percent of banks ran into financial problems. The year 1989 saw 1.8 percent of banks fail.

It is logical to assume that the upturn of the energy consumption cycle coincides with industrial and economic growth. Some evidence for this comes from the construction of subway networks in cities worldwide. Looking at the inauguration date of the very first subway line in every city, one can find a coordination in the time of their appearance in cities around the world.[4] They seem to come in clusters, coalescing into three groups. The first group comprises two cities only, London (1863) and New York (1868). The second group numbers eighteen cities, and construction peaks in 1920, about fifty-six years after the first group. There was a slowing down during the 1940s, but subway inaugurations picked up again in the late 1950s. Construction in the third group is not yet complete, but its cycle peaked in the late 1970s, not far from fifty-six years after the second group. Today there is a declining rate of new cities inaugurating subways, but using the energy oscillation as a model, one can "clock" a fourth wave that will start sometime in the early twenty-first century and peak around 2030.

Peaks in energy consumption coincide with peaks in launching subway projects, both signs of economic prosperity. Such booms make regular appearances every fifty-six years and are not limited to the United States alone. Their universal character is demonstrated by the worldwide succession of primary energy sources already discussed in Chapter Seven.

The competitive substitution of primary energy sources is reproduced at the bottom of Figure 8.3 with one more energy source, work animals, introduced between wood and coal. Nebojsa Nakicenovic has included animals' contribution to the energy supply in terms of the energy content of work-animal feed. Even though his calculations are based on U.S. data only (no worldwide date exist on work-animal feed), I included his animal feed curve here in order to show that there was a primary energy source that peaked around 1870.[5] With the peaks of coal around 1920 and oil around 1975, we find that the rhythm of primary energy substitution is resonating with the same oscillation found in U.S. energy consumption. But the plot is thickening. The price of energy flares up in the same rhythm.

Price indices from different sources are shown at the top of Figure 8.3. To account for inflation, prices have been normalized to 1926 dollars for the fuel and lighting data, and to 1987 dollars for oil. Huge spikes stand out at the beginning of each downturn of the energy cycle. These spikes are so pronounced compared with the usual day-to-day price fluctuations and are so regularly spaced that they inspire confidence in setting forth some daring forecasts. For example, neither price-fixing nor hostilities among the oil-producing countries should succeed in raising oil prices to the 1981 levels for longer than a few months at a time. The price of oil will stay around $20 to $25 a barrel well beyond the year 2000. Energy price are due to hit a high again around 2030 and will mostly reflect the price of natural gas, which will be the dominant energy source of the time.

Cesare Marchetti noted that periods of increasing energy consumption coincide with economic growth leading to a boom. This growth is preceded, and probably caused by, periods during which basic innovations abound. Innovations are like flower bulbs that sprout and bloom

---

FIGURE 8.3   Annual averages for the price of energy are obtained from two different sources for the two periods indicated.* The substitution of energies at the bottom is obtained from Figure 7.6, with the addition of animal feed for the U.S. (intermittent line). The small black triangles point at the peak of each primary energy source's dominance. These points in time coincide with price flares and the beginning of economic decline.

* Fuel and lighting prices from *Historical Statistics of the United States, Colonial Times to 1970,* vols. 1 and 2, Bureau of the Census, Washington, D.C. Oil process from World Energy Outlook, Chevron Corporation, Corporate Planning and Analysis, San Francisco, CA.

ENERGY PRICES, CONSUMPTION, AND SUBSTITUTION, ALL SYNCHRONIZED

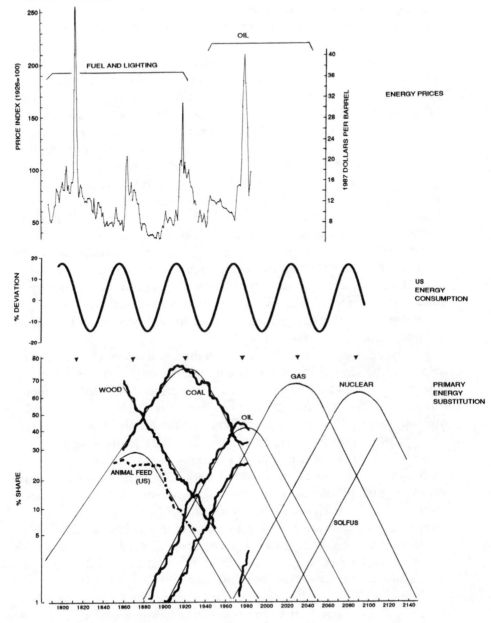

into industries during the boom. To check this further, he looked at the growth of the automobile in the world market and found an extreme regularity in the way populations of cars grew to fill their niches. Like species, they followed natural-growth curves to reach ceilings calculable ahead of time. Moreover, latecomers, such as Japan, grew faster so that in the end they would all reach their saturation level at about the same time: the mid-1990s. In the major Western countries, registered cars have practically completed their niche, just like the Italian niche discussed in the beginning of Chapter Three. A general saturation of the growth of automobile populations across Europe coincides with an economic recession.

Marchetti fitted growth curves on car populations back in 1981.[6] His estimates of the ceilings for most Western countries have been reached already and in some cases surpassed. For example, the registered cars in Italy during 1987 numbered 22.8 million, 10 percent higher than Marchetti's original estimate. This does not necessarily mean he underestimated the Italian market niche. Italy is probably overstuffed with cars these days, and the number may well decrease in the next few years. Such overtaking of the niche size is not unusual. Oscillations are often seen around the saturation level. We will better understand this phenomenon when we make the connection between natural growth and chaos in Chapter Ten.

Oddly, enough, the country in which there is still room left for cars to expand is the United States, where automobiles were first introduced on a large scale. Even though an old-timer, the United States will be among the last countries to be saturated with cars. The ceiling estimate is 173 million;[7] with 137.3 million in 1987 the niche was only 79 percent filled. Yet the 90 percent level will be reached by 1995, the meeting point for saturation of most industrialization processes, as we will see below.

A similar situation is encountered with highways. Construction of highways and paved roads in general can be considered a completed process. The natural-growth curves for the total mileage of roads in ten European countries are reaching the saturation level. With data up to 1970 the perceived ceiling for all paved roads in the United States was estimated as 3.4 million miles. In 1987 this number reached 3.5 million miles, and no more new roads are planned.

Both automobile and road niches are already filled to more than 90 percent. The ceilings of these curves could have been calculated—if with less certainty—back in the 1960s. Therefore, this general saturation cannot be linked to global events such as the oil shocks of the 1970s, or air pollution and other recent environmental concerns of society. Neither is it correct to attribute saturation to economic recession. On the contrary, it is more reasonable to argue that recession results from saturating these niches.

Many of the problems the automobile industry is currently encountering around the world can be explained as follows.[8] When the number of automobiles reaches the saturation level, the industry becomes a supplier of replacements only. Productivity keeps increasing, however, because of competition, tradition, and expected pay increases. A constant level of production coupled with increasing productivity creates redundant personnel and eventual layoffs. Since saturation coincides with recession, so do the layoffs.

It is not only the automobile industry that is in trouble today. Most technological breakthroughs become exhausted more or less at the same time because the cluster of basic innovations born together saturate together. Even if they reach a geographical location late, they usually grow faster there. The simultaneous saturation in many sectors of the economy generates a progressive reduction in employment and low growth in the gross national product—in other words a recession. New growth will follow only after a fresh batch of innovations.

Basic innovations—defined in Chapter Seven as something that starts a new industry or a new kind of product from already existing inventions—occur more readily during recessions, when fewer low-risk investment opportunities are available. These innovations act as catalysts provoking *fundamental changes* in the structure of the economy, its technological base, and many social institutions and relationships. Growth and substitution of old basic technologies by new ones (for example, automobiles for horses, motor for steam, oil for coal, airplanes for automobiles) take place during the upswing of the economic cycle.

We can see in Figure 8.2 that such upswings occurred in 1884–1912 and 1940–1968. We are now in a period of economic recession, a time when innovations abound, but this does not imply an imminent economic upswing. The new activities coming into existence must grow significantly before they have an impact on unemployment and economic development. The energy clock says that we are now approaching the rock bottom of the recession, the mid-1990s. From then onward the rate of growth will progressively increase, but it will only reach a maximum—the next boom—in the early 2020s.

THE WALL STREET CRASH OF 1987

Empirical evidence of long waves in economic activity has existed since the beginning of the Industrial Revolution. William S. Jevons (as cited by

Alfred Kleinknecht[9]) talked about them as early as 1884, and even then he cited earlier works. Much of the debate that has followed this issue concerns whether these variations are accidental or consequences of the way economic growth is linked to profits, employment, innovation, trade, investment, and so forth. A more contemporary scholar, Joseph A. Schumpeter, tried to explain the existence of economic cycles by attributing growth to the fact that major technological innovations come in clusters.[10] An extended list of references on long economic waves can be found in an article by R. Ayres.[11] However, one person deserves special mention, the Russian economist N. D. Kondratieff, whose classic work in 1926 resulted in his name being associated with this phenomenon.[12]

From economic indicators alone Kondratieff deduced an economic cycle with a period of about fifty years. His work was promptly challenged. Critics doubted both the existence of Kondratieff's cycle and the causal explanation suggested by Schumpeter. The postulation ended up being largely ignored by contemporary economists for a variety of reasons. In the final analysis, however, the most significant reason for this rejection may have been the boldness of the conclusions drawn from such ambiguous and imprecise data as monetary and financial indicators.

These indicators, like price tags, are a rather frivolous means of assigning lasting value. Inflation and currency fluctuations due to speculation or politico-economic circumstances can have a large unpredictable effect on monetary indicators. Extreme swings have been observed. For example, Van Gogh died poor, although each of his paintings is worth a fortune today. The number of art works he produced has not changed since his death; counted in dollars, however, it has increased tremendously.

Concerning cycles with a period of fifty-six years I have cited examples in this chapter that are based on physical quantities. Energy consumption, the use of machines, the discovery of stable elements, the succession of primary energy sources and basic innovations have all been reported in their appropriate units and not in relation to their prices. The cycles obtained this way are more trustworthy than Kondratieff's economic cycle. In fact, in the case of primary energy sources, prices indeed followed the same cycle by flaring up at the end of each boom.

Related to money but more physical than prices are bankruptcies and stock market plunges, which manifest themselves during the downward trend of the energy oscillation and hence correspond to a downturn in the economic cycle. Ever since I became aware of the fifty-six-year economic cycle, my concern was not whether a Wall Street crash was around the corner but rather what must one do when faced with an imminent stock market crash. My calculations suggested a crash around 1985, and the minimum precaution to take was to stay away from the stock market.

And so I did. Month after month I resisted the temptation to buy stocks. Colleagues at work would get excited about the bullish market. Favorable terms were offered to buy the company stock. People around me watched their money grow daily. I kept quiet, hoping to be vindicated by the eventual crash—but nothing came. Months went by and the market was still growing. Years went by! Well into 1987 my colleagues had all gotten richer while I was feeling rather sour.

I broke down. It was fall, the leaves were changing color, and I was going to the mountains for the weekend with a friend. I had had enough of holding back. I wanted to be like the others. Friday afternoon I called my bank with an order to buy. I left for the weekend with a feeling that I had finally escaped inaction. I had at last done something, something I would look forward to on Monday.

Over the weekend I enjoyed extraordinary scenery, good weather, reasonable food, and friendship. But there were more important things waiting for me back at work. Monday, October 19, 1987, the stock market crashed. I was crushed. The amount of money I had lost was not so important, but the pain was excruciating. At the same time, on another level, my beliefs had been reinforced. The *system* had behaved according to the plan, as if it had a program, a will, and a clock. I had access to this knowledge early enough. My error was due to human weakness; I had not been scientific. The clock was rather precise, but I should have allowed for an uncertainty of a few percent.

At any rate the crash was over and the stock market largely recovered in a few years. But what remained the same was our general position in the long economic cycle: the recession years. The flares in energy prices in Figure 8.3 can be seen as banners indicating the beginning of an economic downtrend, the end of which we have not yet reached. We will have to wait until 1996 before the growth trend turns around.

MATTERS OF LIFE AND DEATH

There are other human activities that coincide with the fifty-six-year cycle. In Chapter Seven I noted that increases in life expectancy come in waves. While life expectancy is growing all the time, that growth sometimes accelerates and at other times slows. The percentage of deviation around the overall growth trend is seen in Figure 8.4 to correlate with the fifty-six-year oscillation in energy consumption. The peaks in life expectancy closely follow the valleys in energy consumption. The times when life expectancy grows the least come right after a boom, as if to demonstrate that affluence and the decadence often associated with it are not conducive to long life. On the other hand, when recession sets in people tighten their belts, and that has a beneficial effect on life expectancy, whose growth picks up again.

In addition to an adverse effect on life expectancy, the affluence of the boom years appears to have another repercussion: alcoholism. At the center of Figure 8.4 we see the yearly rate of deaths in the United States due to cirrhosis of the liver, which is caused mostly by alcoholism. Its peaks coincide with periods of maximum prosperity, and the simplest explanation for an increase in the number of cases of cirrhosis of the liver is that the number of alcoholics has also increased.

I found another interesting correlation to the energy consumption cycle: the one-mile run. Breaking the record in the one-mile run during the past century and a half has been well documented and holds a secret. Although the running time is steadily getting shorter, sometimes the record is broken year after year—and by a substantial amount—while at other times years go by with no record-breaking or with new records differing at best by only fractions of a second from the previous ones. In Figure 8.4 we can clearly distinguish three such record record-breaking periods fifty-six years apart. Breaking the one-mile-run record is an activity that peaks immediately after a boom. Prosperity and the leisure time associated with it seem to help running. During difficult times, record-breaking becomes rare.

Strangely, breaking the one-mile-run record and increasing life expectancy are two phenomena that oppose each other, even though they both should correlate to good physical condition. When gains in breaking the one-mile record accelerate, increases in life expectancy slow down,

FIGURE 8.4   The graphs on life expectancy and the one-mile-run record have been obtained in a way similar to the energy consumption, namely, as a percentage deviation of the data from a fitted trend. The number of women Nobel laureates is expressed as a percentage of all laureates in a decade.* For the case of cirrhosis the annual mortality is used.

* The graph for women Nobel laureates has been adapted from Theodore Modis, "Competition and Forecasts for Nobel Prize Awards," *Technological Forecasting and Social Change,* vol. 34 (1988): 95–102. Copyright 1988 by Elsevier Science Publishing Co., Inc.

MORE PHENOMENA PULSATING WITH THE ENERGY CONSUMPTION CYCLE

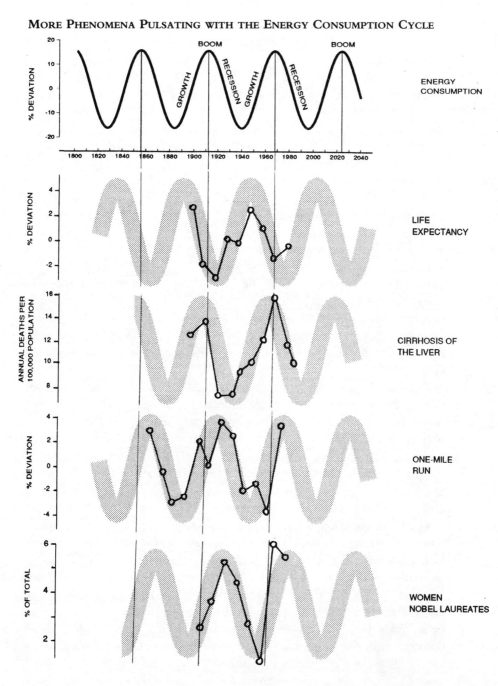

and vice versa. Correlation does not necessarily mean causality, however. It seems unreasonable that life expectancy grows slower because people indulge in running.

It is equally questionable whether it is an economic boom that provokes outbursts of feminism, as may be reflected by an enhanced percentage of women among Nobel prize laureates. In Figure 8.4 we also see the percentage of women Nobel laureates oscillating with time. The period is compatible with fifty-six years. The two peaks occur around 1930 and 1980, coinciding with outbursts of feminism and trailing the corresponding booms. According to the energy clock, prosperity will reach a maximum again around 2025. Whether or not it stems from economic prosperity, the next rise of feminism should also be expected sometime well into the twenty-first century.

The energy consumption clock does not tick only for happy events. Society is anthropomorphic in its moods. In that respect it resembles its constituents, the people. It can be happy, depressed, passive, or aggressive, and this is reflected in the behavior of people who act out roles expressing the current mood of their society. It should be no surprise then to see that homicides are at their peak during desperate times and that homicide rates have gone up and down in tune with the fifty-six-year cycle.[13]

Figure 8.5 shows that homicides in the United States reach their low during the growth years, the rate decreasing by a factor of two between recession and growth years. Even more pronounced—a factor of three between peak and valley—is the shift in preference for the murder weapon. The ratio of firearms to knives oscillates with practically the same frequency. Guns become prevalent toward the end of a boom. Stabbings become prevalent during recessions and reach their peak at the end of the dark days.

Finally, there is also a form of sexual discrimination resonating with the same frequency, the ratio of male to female victims in homicides in the United States. There is a pronounced tendency to kill men—mostly stabbings—when times are tough and recession becomes more and more of a reality. Women are a more frequent target of killings—mostly shootings—during the growth years. The number of female victims reaches a peak just as we enter a boom. Projections indicate that by 1995 annual crime rates in the United States will reflect those of a depression: homicides high (seven per one hundred thousand), mostly male victims (three to one), with increased preference for the knife (40 percent).

All the cases we have seen in this chapter point to recurrent patterns of

MORBID ACTIVITIES ALSO RESONATE WITH THE SAME RHYTHM

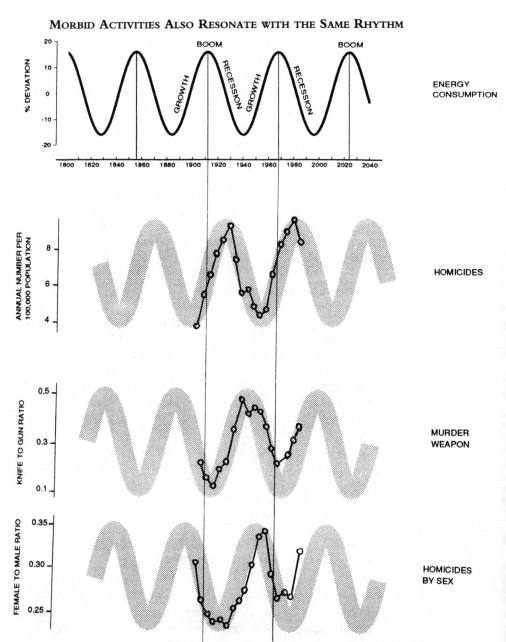

FIGURE 8.5   Below the energy consumption clock, we see the annual rate of homicides. In the middle, the ratio of knife and gun as murder weapon, and on the bottom the ratio of women to men as the victim.

behavior in approximately fifty-six-year intervals. I am convinced that more such patterns could be found if one became sensitive to this underlying pulsation; for example, the recent revival of anti-Semitism through neo-Nazi activities in Europe can be seen as a fifty-six-year echo of the late 1930s. But using only the patterns of behavior that have been charted in this chapter, I might venture to paint a picture of the behavior of society as it is rocked by the fifty-six-year wave. The sequence of events is disturbing, if familiar, even though chronological succession may not necessarily be indicative of a causal relationship:

> During periods of economic growth, criminality is low and so are bank failures. People seem to be busy working, building, increasing their prosperity. As we enter the boom years women become the preferred target for murder. Living through the prosperous years sees people starting to like guns, drink more alcohol, and break running records. Affluence by now does more harm than good, for life expectancy hits a low in its improvement. At the end of the boom, guns are ten times more popular than knives as murder weapons. Right afterward, feminism flourishes, snugly situated between the end of prosperity and the beginning of recession. At this time energy prices flare up like fireworks signaling the end of fun and games. Well into the recession bank failures soar and so do murders. Killings reach a maximum as life becomes difficult; the targets are men this time, and the killers are gradually developing a taste for the knife. There is no bad without some good in it, however. By the end of the recession, life expectancy shows the highest gains, even if competitive sports have suffered. Technological discoveries and innovations abound again, while criminality decreases. The overall sobering up of society serves as a natural preparation for the next growth phase lying ahead. And thus the cycle begins to repeat itself.

EPITAPHS

An explanation can be seen as an attempt to describe a phenomenon by using only knowledge that existed beforehand. Economists indulge in this activity insatiably. Their explanations invariably come after the fact and cannot be tested—"Much like epitaphs," remarks Marchetti, "which for a scientifically trained mind have a strong stink of cemetery."[14]

Physicists also indulge in explanatory activities. In contrast to economists, however, they do not avoid "predicting" the past. A ballistic analysis on a trajectory segment of a flying shell predicts not only where the shell will land but also where it came from. The data I have presented in this chapter show that a fifty-six-year pulsation is encountered over a broad spectrum of human activities, and as a physicist I cannot resist suggesting explanations for this cycle.

Nobel laureate physicist Simon van der Meer[15] pointed out to me while we were discussing these observations that a period of fifty-six years is close to the length of time an individual actively influences the environment. Being an expert in random processes, he suggested that individuals could be acting as fixed "delays" in a never-ending flux of change. In society there are many feedback loops, and despite a continuous arrival of individuals, the existence of a fixed delay—triggered by an instability—could produce "bunching" phenomena with a characteristic period equal to the delay.

Another possible explanation of the regular fifty-six-year oscillation could be linked to periodic phenomena of comparable time scales such as celestial motions. Can we find any cosmic events whose influences on the earth originate far from our planet and pulsate with a beat of about fifty-six years?

Astronomers have long been aware of celestial configurations that recur in cycles. The cycle of Saros, known since antiquity, is based on the fact that identical solar and lunar eclipses occur every eighteen years, eleven days, and eight hours but will not be visible at the same place on the earth. The cycle of Meton, which has been used in the calculation of the date of Easter, is based on the fact that every nineteen years the same lunar phase will occur at approximately the same time of the year. In fact, lunar eclipses recur and are visible at the same place on the earth every 18.61 years. Therefore, the smallest integral year time unit that allows accurate prediction of eclipses at the same place is nineteen plus nineteen plus eighteen, a total of fifty-six years. This fact, as we will see below, played an important role in the construction of Stonehenge.[16]

Lunar and solar eclipses figure prominently in superstition, but their importance goes beyond that. Biological effects, including strange animal behavior, have been observed during eclipses. The mere effect of having the three celestial bodies on a straight line provokes exceptional tides. There are periodicities of fifty-six years on the prediction of tides.[17] But the fifty-six-year period concerns not only eclipses and the alignment of the earth, moon, and sun on a straight line. *Any* configuration of these three bodies will be repeated identically every fifty-six years. Possible effects on the earth linked to a particular geometrical configuration will vary with the fifty-six-year period.

There is one more astral phenomenon, completely independent, that displays a similar periodicity: sunspot activity. For centuries astronomers have been studying spots on the surface of the sun. Cooler than the rest of the sun's surface, these spots can last from a few days to many months, releasing into space huge bursts of energy and streams of charged particles. The effects of sunspot activity are varied and continue to be the subject of scientific study. What is known is that they are of electromagnetic nature and that they perturb the earth's magnetic field, the ionosphere, and the amount of cosmic radiation falling on earth from outer space.

We also know that there is a regular eleven-year variation in sunspot intensity. Sunspot activity reaches a maximum again around 1991, as it did in 1980, 1969, and 1957. During maximum activity, the overall solar output increases by a few tenths of 1 percent. The corresponding temperature change on the earth may be too small to be felt, but meteorologists in the National Climate Analysis Center have incorporated the solar cycle into their computer algorithms for the monthly and ninety-day seasonal forecasts. Every fifth period (five times eleven equals fifty-five years) the timing of the sunspot variation will be close enough to resonate with the fifty-six-year cycle. Moreover, in the three-hundred-year-long history of documented sunspot activity, we can detect relative peaks in the number of sunspots every fifth period (Appendix C, Figure 8.2).

All these celestial influences are probably too weak to significantly affect humans directly. They may affect the climate or the environment, however. There are regularly spaced steps on the continental shelf, and darker circles have been observed on centenary tree cross-sections, with a comparable periodicity.[18] Both indicate environmental changes, and if the environment is modulated by such a pulsation, it is not unreasonable to suppose that human activities follow suit. In fact, observations have linked climatic changes and human affairs.[19]

Men and women of all ages have been fascinated by cyclical phenomena and have looked for them in all disciplines, from astrology to science and from ecology to economics. Most recently their existence has been doubted and even denied, not necessarily for lack of proof but because of a refusal to accept the degree of predetermination ingrained in such a concept.

Does our planet pulse to a fifty-six-year beat? It is interesting to note that far in the dark past, around 2698 B.C., on the other side of the Earth, the Chinese Emperor Huang Ti organized the Chinese calendar in a sixty-year cycle. He set his society's pace to this rhythm, defying the obvious shorter lunar and solar cycles often used for calendric purposes. In so doing he displayed a certain wisdom. Whether or not his people would have been better off with another calendar is an open question. At any rate, to its credit the calendar has been used for the longest period ever. It is still in use today.

Recently a colleague gave me a book, *Stonehenge Decoded* by Gerald Hawkins, which claimed that the architects of Stonehenge possessed an amazing knowledge of the periodic movements of the moon, Earth, and sun. There are fifty-six holes, the so-called Aubrey holes, equally spaced in a circle around Stonehenge. In *Stonehenge Decoded,* Hawkins explains:

> It has always been obvious that the Aubrey holes were important: They were carefully spaced and deeply dug; they served, sporadically, the sacred purpose of tombs; filled with white chalk, they must have been compelling spectacles. But they never held stones, or posts—and, being so numerous and so evenly spaced, they could hardly have been useful as sighting points. What was their purpose?
>
> I think that I have found the answer.
>
> I believe that the fifty-six Aubrey holes served as a computer. By using them to count the years, the Stonehenge priests could have kept accurate track of the moon, and so have predicted danger periods for the most spectacular eclipses of the moon and the sun. In fact, the Aubrey circle could have been used to predict many celestial events.[20]

The number fifty-six is the only integer which would ensure that lunar phenomena repeated accurately over Stonehenge for several centuries.

If Hawkins's postulations are true, the people of Stonehenge, which was constructed sometime between 1600 and 2000 B.C., through observation of celestial phenomena, had become conscious of a natural clock whose hands make a complete circuit every fifty-six years, ticking away periodic influences on life on earth.

Long before I knew of the fifty-six-year cycle I had been intrigued with the possibility that Stonehenge hides secrets. But the mystique around this monument dissipated once I came across Hawkins's explanation of the Aubrey holes and realized that I had something in common with those prehistoric people. Of course, calendars and stone monuments do not illuminate the causes of the cyclical nature of the human activities charted in this chapter, but all these observations have raised my awareness so that I can now easily recognize events that have occurred, and will likely recur, at fifty- to sixty-year intervals.

9

Reaching the Ceiling Everywhere

The first thing physicists do after collecting data is to observe them even before any attempt to analyze or understand them. A great many hidden truths can be inferred, guessed, or at least suggested by just looking at the very first simple representation of the data collected. The usefulness of such an approach in everyday life is often underestimated. The successful repair of a machine or some other piece of equipment that has broken down often consists of little more than a thorough inspection.

Let us observe, then, the data we have accumulated so far in this book. We have seen that man-made objects grow in size, quantity, or market share along S-curves to eventually reach their saturation level. These curves represent natural-growth processes and thus can be extrapolated to time periods for which data do not exist. Moreover, we have also seen a fifty-six-year pulsation modulating many human activities and, in particular, the evolution of the economy. What can we learn by a thorough inspection of all these data? Can their understanding and interpretation be of any practical use in evaluating past trends and predicting future ones? I believe they can. Natural-growth processes, such as the construction of highways and the use of automobiles, along with their relationship to the fifty-six-year cycle, can demystify the persisting present economic recession.

It was mentioned earlier that car populations will saturate their niche in society for most European countries, Japan and, to a large extent, the

United States by 1995. As for the construction of paved roads and high-ways, the main effort will be in maintaining and improving what already exists rather than adding more. Road construction in the United States enjoyed a surprising comeback in the early 1980s but it should now cease. Contrary to popular belief, paved roads "diffused" into society *before* cars; they reached the 90 percent level about a decade earlier, despite the fact that all these processes merged together as they saturated.

In general, one could say that society has achieved the capacity that satisfies its needs for this means of transportation and is reluctant to invest further. We are at the end of the era in which people were preoccupied by the automobile. After all, air travel has been gaining importance relative to road travel ever since 1960 when the latter enjoyed the lion's share of the transportation market (see Chapter Seven).

In 1960 the automobile was at its zenith. In the division of U.S. intercity passenger traffic among trains, cars, and airplanes shown in Figure 9.1, the percentage of traffic attributed to trains (buses were included in this category) has been systematically declining, while that of airplanes has been rising. The share of cars rose until 1960, reaching close to 90 percent, but then it began to decline. Cesare Marchetti believes that the automobile, in spite of its dominant position at the time, "felt" the rising threat of airplanes.[1] He whimsically adds that at the moment when the automobile's market share started declining, cars were "masquerading themselves as airplanes with Mach 0.8 aerodynamics, ailerons, tails, and 'cockpit' instrument panels. The competitor is the devil—in this case the airplane—and as peasants still do in the Austrian Alps, to scare the devil, one has to dress like the devil."

Even though airplanes are displacing cars percentage-wise in the above example, the growth of air traffic in absolute terms seems to be slowing down. A graph of the ton-kilometers carried every year worldwide shows that air traffic is heading for a point of saturation toward the end of the century. The S-curve that best fits the data has a ceiling at 360 billion ton-kilometers per year which makes that niche 90 percent complete by the year 2000 (Appendix C, Figure 9.1).

The agreement between data and curve turns out to be remarkable, considering that there have been at least two sharp fuel price increases during this period, one in 1974 and another in 1981. One may have expected that the diffusion of air traffic was impacted by the price of fuel on which it depends so directly. Not at all! The *system* seems to com-pensate internally for such events so as not to deviate from its natural course.

As we head for the mid-1990s, a multitude of growth curves are

DRESSING UP AS THE DEVIL

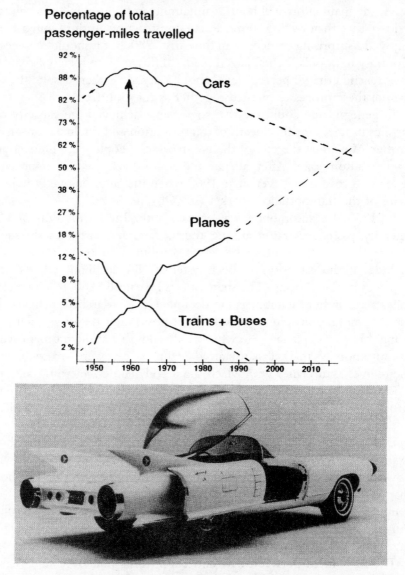

FIGURE 9.1   On the top we see the division among means of transportation competing for intercity passenger traffic in the United States. The vertical scale is logistic. The intermittent lines are extrapolations of fits on the data shown. On the bottom, a 1959 Cadillac Cyclone is representative of car "behavior" at the moment when their dominance started being challenged by airplanes. Adapted from work carried out at the International Institute of Advanced Systems Analysis; photograph from General Motors.

slowing down, reaching a ceiling. For automobile populations and high-way construction, the saturations are understandable; there is no need for further growth. In other cases, such as air traffic, the slowdown must be considered as only the end of a phase because the airway infrastructure is young, and more growth phases must follow.

The group at IIASA (Cesare Marchetti, Nebojsa Nakicenovic, and Arnulf Grubler) has indulged in fitting logistic functions on growth processes by the hundreds for over a decade. The cases span the last three hundred years and predominantly concern the construction of infrastructures and the substitution and diffusion of technologies. When Alain Debecker and I came across Arnulf Grubler's book, *The Rise and Fall of Infrastructures,* which contains many of the IIASA results, we decided to produce one drawing on which all these curves were superimposed. The original fits had been reported as straight lines on logistic scales. We transformed them back into S-shapes. In Figure 9.2, we included some of the curves that we had determined ourselves concerning computer innovation, subway inaugurations, and airways.

There are more than fifty growth processes shown, each with its ceiling normalized to 100 percent. In the table below, The Timing of Technological Growth Curves, the name of each technology is listed in chronological order along a line at 50 percent. As we expected, the curves coalesced into clusters, merging together as they approach the ceiling that indicates saturation. The left-most three curves, in the late eighteenth and early nineteenth centuries, represent canal construction. The next group includes primarily diffusion of steamships and railway networks; the fast-rising ones to their right are automobile substitutions of horses. In the twentieth-century cluster at the far right, road construction and replacement of steam locomotives are seen to grow slowly, followed by automobile populations, which grow more quickly. Among the recent curves are some that describe innovation in the computer industry.

We then compared these natural-growth curves to the fifty-six-year cycle of energy consumption, which coincides with the economic cycle. We observed a remarkable correlation between the time these growth curves approach their ceiling and the valleys of the economic cycle. It thus becomes evident that boom or prosperity, which occurs during a peak of the cycle, coincides with technological growth while recession coincides with saturation of these technologies. The industries built around these technologies follow the same curves and also saturate at the same time.

**WHEN MANY TECHNOLOGICAL PROCESSES REACH SATURATION TOGETHER,
THEY PRODUCE A RECESSION**

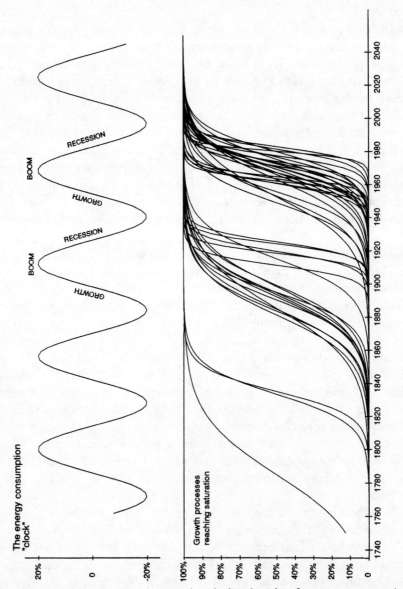

FIGURE 9.2   At the top we see again the idealized cycle of energy consumption
as obtained in Chapter 8. At the bottom we see S-curves describing technological growth and substitution processes all normalized to reach 100% when
they are completed. The rising part of most curves coincides with the upswing
of the cycle indicating growth and prosperity, while "coordinated" saturation
coincides with economic recession.

Three waves of industrial growth are outlined by the bundling of the curves. The times when the curves merge indicate barriers to economic development. Growth processes do not traverse them. Exceptionally, a process may "tunnel through" the economic recession and continue growing into the next cycle. A few examples are mentioned in the table. These S-curves stretch across the blank areas between the three bundles of lines in Figure 9.2 (Appendix C, Figure 9.2). Between the first wave (rather poorly defined because of insufficient cases) and the second wave, it was canal construction in Russia that continued developing. Between the second and the third waves, the traversing process was the construction of oil pipelines in the United States. Oil technology kept growing, unaffected by the recession of the 1930s, a period dominated by steam/coal technology. The pipeline projects contributed significantly to the recovery from that major recession.

During the present recession, the processes that will continue growing and thus contributing to the economic recovery are the buildup of airway connections between American cities and pollution abatement (to be discussed later in this chapter). To a lesser extent natural gas pipelines, nuclear energy power stations, and the computer industry will also contribute to the recovery of the economy. On a worldwide scale, among the industrialization activities that will still be incomplete by the end of the century are road construction in the U.S.S.R. and car populations in Canada and New Zealand.

It should also be noted that the clustering of the saturation levels of the various industries and the technologies behind them is ultimately related to the clustering of basic innovations mentioned in the last chapter. While the appearance of basic innovations leads to growth and prosperity, innovations that come into being together become exhausted together, producing low growth, reduction in employment, and increased competitiveness.

In today's recession the aging of the old innovations throws a veil of temporariness on enterprising endeavors. Products are going through progressively shorter life cycles in a desperate quest for differentiation when the technologies at hand no longer justify it. The lifetimes of companies themselves are being reduced, with a growing number of mergers and buy-outs. (Even the cycle of personal relationships is getting shorter; the average duration of relationships, including marriages, has been reported to be less than 2.5 years in the United States.) Social living in general, seen from all angles, is getting ever more competitive and demanding. Succeeding or even simply surviving in today's society is

THE TIMING OF TECHNOLOGICAL GROWTH CURVES

Unless otherwise indicated (as for the subway-related and the computer-related graphs), the dates have been read off various drawings in Grubler's *The Rise and Fall of Infrastructures*. Tunneling-through processes are not shown in Figure 9.2 (representative ones are shown in Appendix C, Figure 9.2).

	Level of Saturation		
	10%	50%	90%
First Industrial Growth Wave			
Canal construction—England	1745	1785	1824
Canal construction—France	1813	1833	1853
Canal construction—United States	1819	1835	1851
Tunneling-through process:			
Canal construction—Russia (first			
burst)	1781	1837	1892
Second Industrial Growth Wave			
Railway construction—Germany	1853	1881	1909
Railway construction—Austria	1858	1884	1909
Substituting steam for			
sails—United Kingdom	1857	1885	1912
Substituting steam of sails—United			
States	1849	1885	1921
Substituting steam for			
sails—Austria/Hungary	1867	1890	1914
Railway construction—United			
States	1865	1892	1919
Railway construction—Worldwide	1866	1893	1921
Substituting steam for			
sails—Worldwide	1868	1894	1921
Substituting steam for			
sails—Germany	1874	1895	1915
Substituting steam for			
sails—France	1868	1898	1928
Substituting steam for sails—Russia	1872	1899	1926
Substituting cars for			
horses—France	1903	1911	1919
First wave of subway			
inaugurations—Worldwide[2]	1892	1915	1938
Substituting cars for			
horses—United States	1911	1918	1925

	Level of Saturation		
	10%	50%	90%
Substituting cars for horses—United Kingdom	1910	1919	1929
Substituting cars for horses—Worldwide	1916	1924	1932
Substituting cars for horses (freight)—France	1916	1925	1934
Tunneling-through process:			
Oil pipeline construction—United States	1907	1937	1967
Third Industrial Growth Wave			
Railway construction—U.S.S.R.	1928	1949	1970
Growth of paved roads—United States	1921	1952	1984
Substituting motor for steamships—United Kingdom	1928	1953	1978
Canal construction—U.S.S.R. (second burst)	1936	1956	1975
Phasing out of steam locomotives—Germany	1950	1961	1971
Phasing out of steam locomotives—Austria	1952	1961	1970
Phasing out of steam locomotives—U.S.S.R.	1956	1962	1968
Natural gas pipeline construction—United States	1936	1963	1990
Phasing out of steam locomotives—France	1954	1963	1973
Phasing out of steam locomotives—United Kingdom	1958	1965	1971
Car population—United States	1942	1967	1995
Jet engine performance—Worldwide	1952	1967	1982
Car population—United States	1946	1967	1989
Natural gas pipeline construction—Worldwide	1948	1969	1990
Passenger aircraft performance—Worldwide	1953	1970	1988

	Level of Saturation		
	10%	50%	90%
Car population—Australia	1954	1970	1987
Car population—Italy	1960	1971	1982
Car population—Worldwide	1957	1972	1988
Car population—Austria	1960	1973	1986
Car population—Spain	1964	1974	1983
Second wave of subway inaugurations—Worldwide	1955	1975	1995
Passenger air traffic—Worldwide	1962	1976	1991
Car population—Japan	1969	1977	1986
Total air traffic—Worldwide	1962	1980	1997
Percentage of cars with catalytic converters—United States	1975	1980	1985
Personal computer manufacturers—Worldwide[3]	1972	1981	1991
Personal computer models—Worldwide	1975	1983	1991
Tunneling-through processes: Airways—United States[4]	1963	2001	2039
Paved roads—U.S.S.R.	1949	1980	2012
Computer models (excluding personal computers)—Worldwide	1973	1988	2002
Computer manufacturers (excluding personal computers)—Worldwide	1976	1991	2006

becoming harder all the time. Naturally enough, the question on everyone's mind is, "Just how long will these conditions last?" To answer that question we can look to natural-growth curves and the periodic economic cycle to which they are related.

Not long ago I was showing my observations of cyclical human behavior to Michael Royston, a friend who had been teaching environmental sciences in the International Management Institute of Geneva for a number of years. Royston became excited and quickly produced an unpublished paper of his own, written in 1982, in which he talked about the same fifty-six-year cycle but from another angle.[5]

Royston's thesis was that life progresses in spirals and that long-term growth follows a spiral which passes successively through four phases:

discharge, relaxation, charge, and tension, after which it returns to the starting point, but enriched with new knowledge, experience, and strength. The period of discharge is characterized by economic prosperity or a boom; relaxation is characterized by recession; charge is a period of new order and new technology; and tension is a time of growth leading once again to discharge and boom. The entire cycle is completed and repeated in fifty-six-year intervals. In the most recently completed cycle, reproduced in Figure 9.3, Royston considers the years 1912 and 1968 as high points of affluence, followed by periods when the world tumbled into war and economic turmoil. In the spiral that begins with 1968 there would appear to be an ominous echo of 1940 in 1996.

Royston made an attempt to connect his fifty-six-year cycle with environmental issues, as well as with human behavior:

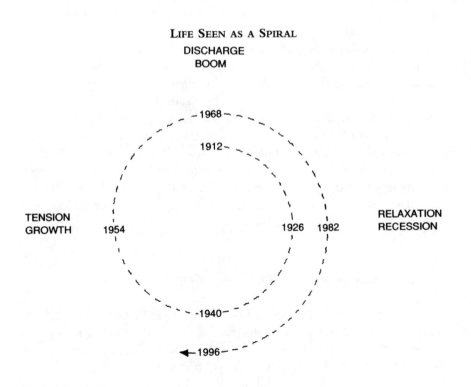

LIFE SEEN AS A SPIRAL
DISCHARGE
BOOM

1968

1912

TENSION
GROWTH
1954

RELAXATION
RECESSION

1926   1982

-1940-

1996

CHARGE
NEW ORDER, NEW TECHNOLOGY

FIGURE 9.3   The Royston spiral.

Humans spend the first twenty-eight years of their lives acquiring or "charging," first an affective, then a physical, then an intellectual, and finally a spiritual capability, each building successively on a seven-year spiral. The second twenty-eight years see the human in a state of "tension" as parent, contributor to society, thinker. The final twenty-eight years the person becomes "discharged" affectively and spiritually, reaching the full age of three times twenty-eight to relax in eternity.

Royston rolled back history and pointed out several events that acted as historical turning points, all occurring very close to fifty-six years apart. His list included the invention of the floating compass (1324), the invention of gun powder and gun making (1380), the invention of the printing press (1436), the discovery of America (1492), the beginning of the Reformation (Luther and Calvin, 1548), the defeat of the Spanish and the rise of the Dutch (1604), the arrival on France's throne of Louis XIV (1660), the rise of the English Empire (1716), and the American War of Independence (1772).

The fifty-six-year periods that followed these events saw successive transfers of power, from the French to the British with the end of the Napoleonic era (1828–1884), from the British to the Germans with the new technologies of chemicals, automobiles, airplanes, and electric power (1884–1940), and from the Germans to the Americans with such new technologies as plastics, transistors, antibiotics, organic pesticides, jet engines, and nuclear power (1940–1996). Post-1992 Europe, aided by the collapse of Communism, can conceivably wrestle power away from the Americans in the period 1996–2052. It is also during this period that the current recession will bottom out, and new technology and growth will lead to more prosperous times.

CLOCKING THE TRANSPORT OF THE FUTURE

One may argue that Royston's work is subjective; that he is choosing to look only at those historical events that suit his theory—there is so much in history that one may always find something interesting around a given date. One cannot argue the same way, however, against the work of Nakicenovic on the U.K. Wholesale Price Index documented since the sixteenth century. In Figure 9.4 we see the variation of this index, smoothed with an averaging technique, eloquently pointing out a long economic wave with mean periodicity of 55.5 years.

Both Royston and Nakicenovic corroborate the more quantitative

A Periodic Oscillation Recorded over Five Centuries

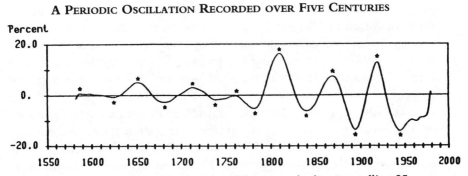

FIGURE 9.4   The U.K. Wholesale Price Index smoothed over a rolling 25-year period with respect to a 50-year moving average. This procedure washes out small fluctuations and reveals a wave. The periodicity turns out to be 55.5 years.*

results on the fifty-six-year cycle presented in Chapter Eight. All these observations underline the fact that reaching the ceiling everywhere today does not mean we are heading toward a final end. We are simply reaching the bottom of the valley between two wave crests. The long economic wave that often carries the name of N. D. Kondratieff, the Russian economist who first talked about it, can be thought of as a chain of bell-shaped life cycle curves. Each life cycle can be associated to an S-curve that extends over half a century and serves as an envelope containing many other processes. Growth generally comes to a standstill— saturation—when its rate reaches a low point halfway between crests.

We are now in such a low growth period between crests, but saturation, competitiveness, and recession will not keep tightening their noose forever. The mid-1990s will be a period of discontinuity, characterized by exhaustion of the old development path and the transition of a new one. In order to define the new path, innovation and fundamental restructuring become necessary. Profound social and institutional changes may be inevitable. The general saturation reached every fifty-six years is the Kondratieff "barrier" that stops most of the old ways and makes place for new ones. Once that process is under way we will begin to see a period of growth and increasing prosperity.

We can look again at the history of transport in the United States for

* Adapted from a graph by Nebojsa Nakicenovic in "Dynamics of Change and Long Waves," report WP-88-074, June 1988, International Institute of Advanced Systems Analysis, Laxenburg, Austria. From the article by R. Ayres, "Technological Transformations and Long Waves," parts I and II, *Technological Forecasting and Social Change*, vol. 37, nos. 1 and 2 (1990). Copyright 1990 by Elsevier Science Publishing. Reprinted by permission of the publisher.

a view of this process in action. The substitution of transport infrastructures, discussed in Chapter Seven, showed that canals gave way to railways, which in turn were substituted by roads that are now yielding to airways. The data representing the length of each infrastructure can be graphed as a succession of normalized S-curves, each one going to 100 percent when the final mileage is reached. In Figure 9.5, originally drawn by Nakicenovic, I have added the diffusion of airways and a future means of transportation called Maglev (magnetic levitation train). For the airways I used an estimate of 3.2 million miles as a final ceiling.[6] The projection says that the completion of the airway network is not due until the middle of the twenty-first century, well beyond the Kondratieff barrier of 1995. By the year 2000 the total mileage of air routes connecting United States cities will be more than 2.5 times that of 1980.

Air transport would appear to be an exception to the exhaustion of

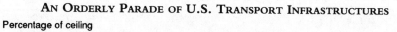

AN ORDERLY PARADE OF U.S. TRANSPORT INFRASTRUCTURES

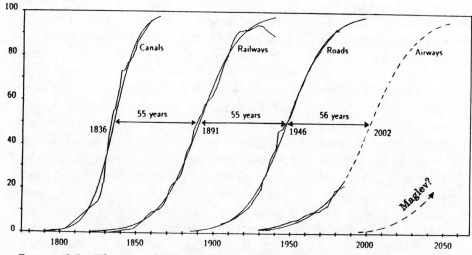

FIGURE 9.5   The growth in length of each infrastructure is expressed as a percentage of its final ceiling. The absolute levels of these ceilings in miles are quite different (see text). For airways the ceiling has been estimated. The 50 percent levels of these growth processes are regularly spaced 55 to 56 years apart. A future infrastructure (called Maglev) may start sometime around the turn of the century, but its halfway point should be rather close to 2058.*

* Adapted from a graph by Arnulf Grubler in *The Rise and Fall of Transport Infrastructures*, (Heidelberg: Physica-Verlag, 1990), excluding the lines labeled "Airways" and "Maglev?" Reprinted by permission of the publisher.

technologies typical of periods of recession, contrary to the evidence provided earlier that world air traffic was nearing its saturation level. Here we see that the building up of air traffic cannot cease soon. Another growth curve must be anticipated starting around the turn of the century. One more reason to expect a new growth curve is the evolution of a *new* technology. This second aviation wave will most likely involve the development of supersonic airplanes, probably using liquid hydrogen for fuel, produced with the help of nuclear reactors. Such a link between a transport system and a primary energy source has been seen before (for example, steam with coal, cars with oil) and should have an impact on the evolution of nuclear energy itself.[7]

There are two other observations to be noted in Figure 9.5. One is that all transport infrastructures follow similar trajectories to reach their ceiling, even if these ceilings are widely different:

Infrastructure	*Ceiling (total mileage)*
Canals	4,000 miles
Railways	300,000 miles
Roads	3.4 million miles
Airways	3.2 million miles (estimated)

The second observation is that the 50 percent points in the growth of each infrastructure are about fifty-six years apart, in resonance with the cycle first mentioned in Chapter Eight. Each transportation system is intimately linked to a primary energy source: canals to animal feed, railways to coal, and cars to oil. The appearance of the new energy source precedes the saturation phase of the current transportation system. According to this line of thought, the airplane fuel of the future will not be gasoline! Airplanes are using gas just as locomotives used wood in the early decades of railroads (at least up to the 1870s), even though coal was growing in importance as a primary energy source. The newest primary energy source today is nuclear, and that is what argues for the development of a nuclear-hydrogen system to power the major part of the airway cycle.

Thus, we see that new transport infrastructures are conceived during recessions, with the emergence of new innovations and technologies, but they grow slowly. Their maximum rate of growth—the halfway point—occurs during (and contributes to) the economic growth of the following cycle. Railways came into use in the late 1820s, and peak activity in railway construction contributed to the recovery from the broad recession of the 1880s. Paved roads began to be built in the 1880s and contributed mainly to the economic recovery of the 1930s, a time when air transport was being undertaken. The maximum rate in the buildup of air transport will help the recovery from the economic low of the mid-1990s.

In extrapolating the same pattern we can estimate that a future means of transportation, labeled Maglev in the figure, should be undertaken around 1995, enter the market significantly around 2025, by claiming a 1 percent share of the total length in all transport infrastructures, and reach its maximum rate of growth close to 2058, thus helping the recovery from yet another depression in 2050. In keeping with the usual pattern, this new transport infrastructure must provide a factor-of-ten improvement in speed or, more precisely, in productivity (load times speed.) A supersonic plane based on new technology—possibly liquid hydrogen—may provide Mach 8 in speed but will be useful only for long distances. Supersonic flight is obviously not the answer for linking cities within a continent. Since speed of existing airplanes is sufficient to serve shuttle links such as New York to Boston, the future vehicle must increase the number of passengers by a factor of ten, the equivalent of the Boeing 757 or the European Airbus, but with a carrying capacity of close to twenty-five hundred passengers! The problems arising from handling that many passengers in one flying vessel would be formidable.

An alternative may be found in Maglevs. These trains have been studied for some time by the Japanese; they move at a mean speed of up to six hundred miles per hour, and from the point of view of speed and cost they are like airplanes. Maglevs should connect only core cities in order to justify their high capacity and investment costs. Considering the head start of the Japanese, the first operational Maglev may be between Tokyo and Osaka, creating an extended city of 100 million around the turn of the century. Maglevs may finally *functionally* fuse other large urban areas, producing "linear supercities as 'necklaces' of old ones."[8]

Maglevs would serve as links between supersonic air routes, and therefore the best location of their stations would be airports. Airports are already becoming communication junctions, a function served by railroad stations at the end of the last century. For the picture to be complete, fast subway connections to the airports are necessary. The subway system, far from being satisfactory today, is going to be the weakest link in the transport system of the future. Fortunately, as forecasted in the previous chapter, a new wave of ambitious city subway projects is to be started around the turn of the century, a time when the current recession will be behind us and we will have embarked on a period of new growth.

FUTURONICS

The natural phenomena mentioned so far in this book can serve as forecasting tools: *invariants* representing equilibrium positions, *S-curves* describing competitive growth and substitutions, and the *fifty-six-year cycle* clocking cyclical variations. Exploiting these notions I am tempted to look further into the future to paint part of the picture of our children's generation.

According to the Kondratieff cycle, we are heading more deeply into a general economic recession but at the same time into a period in which new innovations will show up at an increasing rate. Most of the fundamental discoveries that will lead to new industries at the beginning of the next century have already been made. There have been many inventions in the last few decades, but we must wait to see which ones will lead to industrial successes before we pronounce them basic innovations.

Again according to the fifty-six-year cycle, the period 1996 to 2024 should be a period of growth leading to prosperity not unlike what happened between 1940 and 1968. Real growth will provide opportunities for entrepreneurship, which should succeed the era of bureaucracy from which we have just emerged. Opportunities should surface in a wide range of activities. The individuals best suited for this new period of growth will be multitalented men and women who have been trained in breadth rather than in depth, generalists rather than specialists. They will have an easier time making a niche for themselves as entrepreneurs at the turn of the century. It is the generation after them that will provide the bureaucrats of the 2020s.

Fifty-six years ago industrial and economic growth was greatly stimulated by the need for reconstruction following World War II. Today a cross-industry saturation is once again strangling economic development, fifty-six years after a similar crisis in the thirties. Will it be followed by another large-scale violent conflict around 1995 comparable to that of World War II? That would seem to be a foredoomed forecast these days, in view of the end of the cold war and the collapse of Communism in Europe. Still, a period of rapid growth, such as the one predicted for the turn of the century, is usually preceded by major destruction

This logic for stimulating growth has probably been one factor behind the renewal rituals practiced by earlier societies such as the Aztecs. All Middle American civilizations, including the Mayans with their extraordinary refinement in mathematical and astrological knowledge, used two calendars. A ritual year of 260 days ran in parallel to the solar year of 365 days. The least common multiple of 260 and 365 is 18,980 days, or 52 years. At the end of each 52-year-cycle, the Aztecs performed a rite in which pottery, clothes, and other belongings were voluntarily destroyed, and debts forgotten. A ceremony called *tying up the years* culminated with the renewal-of-the-fire ritual to ensure that the sun will rise again.

Such a voluntary destruction of assets would seem unacceptable to us today. Thus, excluding a major war, it is difficult to imagine a source of destruction for Western society in the mid-1990s that would cause large-scale material damages and trigger the next growth phase of the Kondratieff cycle. Yet the cycle predicts that important economic development and reconstruction efforts should begin later in this decade and span the better part of the following twenty-five years. What, then, will trigger them?

I was at an impasse with the forecast of an imminent worldwide catastrophe until I came across the February 1990 issue of Larry Abraham's *Insider Report,* which directed me toward an unusual book where I found a resolution for my conflict.[9]

On the Possibility and Desirability of Peace

In 1967 a little book was published by Dial Press titled *Report from Iron Mountain on the Possibility and Desirability of Peace.* In the introduction, written by Leonard C. Lewin, we are told that "John Doe," a professor at a large midwestern university, was drafted by a governmental committee in August 1963 "to serve on a commission 'of the highest importance.' Its objective was *to determine, accurately and realistically, the nature of the problems that would confront the United States if and when a condition of 'permanent peace' should arrive, and to draft a program for dealing with this contingency."*[10]

The real identities of John Doe and his collaborators are not revealed, but we are told that this group met and worked regularly for over two and a half years, after which it produced its report. The report was eventually suppressed both by the government committee and by the group itself. Doe, however, after agonizing for months, decided to break with keeping it secret, and he approached his old friend Lewin, asking for help with its publication.

Lewin explains in his introduction why Doe and his associates preferred to remain anonymous and did not want to publicize their work. It was due to the conclusions of their study:

> Lasting peace, while not theoretically impossible, is probably unattainable; even if it could be achieved, it would almost certainly not be in the best interests of a stable society to achieve it.
>
> That is the gist of what they say. Behind their qualified academic language runs this general argument: War fills certain functions essential to the stability of our society; until other ways of filling them are developed, the war system must be maintained—and improved in effectiveness.

In the report itself, after explaining that war served a vital subfunction by being responsible for major expenditures, national solidarity, and a stable internal political structure, Doe goes on to explore the possibilities of what may serve as a substitute for war, inasmuch as the positive aspects of it are concerned. He writes: "Whether the substitute is ritual in nature or functionally substantive, unless it provides a believable life-and-death threat it will not serve the socially organizing function of war."

In Section 6, entitled "Substitutes for the Functions of War," Doe quantifies the economic conditions that must be satisfied:

> Economic surrogates for war must meet two principal criteria. They must be "wasteful," in the common sense of the word, and they must operate outside the normal supply-demand system. A corollary that should be obvious is that the magnitude of the waste must be sufficient to meet the needs of a particular society. An economy as advanced and complex as our own requires the planned average annual destruction of not less than 10 percent of gross national product if it is effectively to fulfill its stabilizing function. When the mass of a balance wheel is inadequate to the power it is intended to control, its effect can be self-defeating, as with a runaway locomotive. The analogy, though crude, is especially apt for the American economy, as our record of cyclical depressions shows. All have taken place during periods of grossly inadequate military spending.

Among the alternatives that the study group considered were war on poverty, space research, and even "the credibility of an out-of-our-world invasion threat." The most realistic, however, may be war on pollution. In Doe's words:

It may be, for instance, that gross pollution of the environment can eventually replace the possibility of mass destruction by nuclear weapons as the principal apparent threat to the survival of the species. Poisoning of the air, and of the principal sources of food and water supply, is already well advanced, and at first glance would seem promising in this respect; it constitutes a threat that can be dealt with only through social organization and political power. But from present indications it will be a generation to a generation and a half before environmental pollution, however severe, will be sufficiently menacing, on a global scale, to offer a possible basis for a solution.

After the publication of the Report, guessing games of "Who is Doe?" and "Is the Report authentic?" started circulating among academic and government circles. The White House conducted an inquiry and concluded that the work was spurious. Then in 1972 Lewin admitted authorship of the entire document. He commented as follows:

What I intended was simply to pose the issue of war and peace in a *provocative* way. To deal with the essential absurdity of the fact that the war system, however much deplored, is nevertheless accepted as part of the necessary order of things. To caricature the bankruptcy of the think tank mentality by pursuing its style of scientistic thinking to its logical ends. And perhaps, with luck, to extend the scope of public discussion of "peace planning" beyond its usual stodgy limits.

While reading Lewin's fictitious report I realized that his prediction on pollution is now proving ominously realistic. One to one and a half generations from the mid-1960s brings us to the 1990s and the subsequent decade. The damage we have done to our environment during this period can be equated to the major destruction caused by war. The billions of dollars that will be required, along with new technologies to replace the old, to rescue the environment can be likened to the vast economic and technological resources necessary to repair the damage caused by a major war. In this sense, then, pollution is proving to be a successful surrogate for war, but at the same time there is some good news. Pollution abatement in the United States today uses around 2 percent of the gross national product. If as a war substitute it must reach 10 percent, the expenditures for pollution abatement growing along a natural-growth curve will "tunnel through" the current economic recession and contribute significantly to the economic recovery that leads to the prosperity predicted in the first quarter of the next century.

The West's favorite candidate for a major war in the 1990s has always been expected from the East, but this notion must now be abandoned as Communism goes through end-of-life spasms. What still remains a credible alternative is an all-out war on pollution. It could be argued that we have not yet marshaled the vast resources necessary to win that war, but it might be noted that Earth Day falls on Lenin's birthday!

Global Village in 2025

Assuming humanity fares well in its war on pollution, we can now look beyond the next decade by using some of the ideas presented in Chapter One—in particular, the concept of invariants. Let us try to picture urban life in the next century.

Invariants are quantities that do not change over time and geography because they represent a natural equilibrium. Such universal constants can be exploited to reveal future shapes of society. Cars, for example, display a curious characteristic in regard to the average speed they can attain. Today in the United States it is about thirty miles per hour, hardly changed since Henry Ford's time.[11] Car users seem to be satisfied with an average speed of thirty miles per hour, and all performance improvements serve merely to counterbalance the time lost in traffic jams and at traffic lights.

This invariant can be combined with another universal constant, the time available for traveling. Yacov Zahavi studied the equilibrium between travel demand and supply, and urban structure.[12] He and his collaborators showed the following: (a) Inhabitants of cities around the world devote about one hour and ten minutes per day to urban transport, be it by car, bus, or subway. (b) For this transport they all spend a rather constant percentage of their income, 13.2 percent in the United States, 13.1 percent in Canada, 11.7 percent in England, and 11.3 percent in Germany. (c) What varies from one individual to another is the distance they cover; the more affluent they are, the further they commute. Zahavi concludes that whenever there is a real discrepancy between these constants and a person's behavior, the natural balance is disturbed and tensions and rejection might ensue.

Coming back to the car, at an average speed of thirty miles per hour, a traveling time of one hour and ten minutes a day translates to about thirty-five miles. This daily "quota" for car mileage is indeed corroborated from data on car statistics.[13] During the last fifty years, yearly mileage in the United States has been narrowly confined to around ninety-five hundred miles, despite the great advances in car speed and acceleration over this period. It turns out to be 36.4 miles per working day, in good agreement with the earlier result.

The long-term stability evidenced by the above invariants is proof of a balance between time available, disposable income, and road networks. The average distance covered per day becomes the natural limiting factor in defining the size of urban areas. Communities grow around their transport systems. If it takes more than seventy minutes to get from one point to another, the two points should not reasonably belong to the same "town." Cars permitted towns to expand. When people traveled only on foot, at three miles per hour, towns consisted of villages not much larger than three miles in diameter.

There was a factor of ten in speed between foot and car transport, but also between car and airplane transport, taking the average airplane speed as around three hundred miles per hour. Airplanes expanded the limits of urban areas further, and it is possible today to work in one city and live in another. Air shuttle services have effectively transformed pairs or groups of cities in the United States, Europe, and Japan into large "towns."

If we now imagine the supersonic airplane of the future with an additional factor of ten, say an average of three thousand miles per hour, we can visualize the whole Western world as one town. In his book *Megatrends,* John Naisbitt claims that Marshall McLuhan's "world village" is being realized today as communication satellites link the world together informationally.[14] This is not quite accurate. Information exchange is a necessary but not a sufficient condition. It is true that empires in antiquity broke up when they grew so large that it took more than two weeks to transmit a message from one end to the other. But it is also true that communications media are poor substitutes for personal contact (see Chapter Six). The condition for a "world village" is that it should take not much more than one hour to physically reach any location.

One way of achieving this is a supersonic air connection among distant cities, with Maglevs acting as high-speed, high-volume connections between closer core cities, and expedient subway access to the airports. There is not much room in this picture for conventional railways. They seem to have no future as a network and will inevitably decline world-wide during the economic cycle starting in 1995. In some fast network branches, such as the French TGV connecting Paris to Lyon, trains may linger for a while, losing ground slowly to improving subsonic jumbo plane service over the same routes.

But what will happen to the car, this longtime favorite means of transportation, toy, and hallmark of individual freedom and mobility? Cars are not suitable for the densely populated cities that are emerging everywhere. Urban centers with a high population density increasingly provide public services that eliminate the need for a personal car. This is becoming more and more evident with expanding areas in city centers in which automobile traffic is excluded. On the other hand, urbanization is still on the rise. World population is imploding exponentially (this trend will eventually slow down) into cities of ever-growing numbers and sizes.[15] Combining the above elements one may conclude that a long time from now the majority of the population will not be using a private car for urban or intercity transport. Possible remaining uses are for hobbies and pleasure. Cesare Marchetti's long-term view of future car employment is inspired by clues already found in the American society: "*Cars* . . . may become small *trucks* carrying voluminous adult toys around the countryside."

The day when cars start declining in number is still far in the future. For the time being the growth of car populations is stopping in most of the Western world, provoking a saturation in the automobile industry. Similar and simultaneous saturations in many other industries give rise to the present economic recession. But the cyclical phenomena seen in Chapter Eight assure us of a timely change—in particular, of a new cluster of innovations to flourish in the 1990s, giving rise to new industries. Moreover, a few of the old industries have not yet reached saturation levels; for example, electronics and computers, natural gas pipelines, nuclear energy power stations, and subway transportation infrastructures. Finally, two growth processes that are still rather young and can be expected to act as the workhorses of the economic development during the next two decades are air route connections and pollution abatement. On a global scale there are many parts of the world where the growth of the old industries is not yet complete; a case in point is road construction in the U.S.S.R. Taken together, all of these factors point to a period in the very near future when economic growth will start increasing again.

10

If I Can, I Want

Logistic functions are well suited to describing natural growth, and if all growth processes fell in this category, one would need little additional mathematics. Alas, there are phenomena that proceed in ways so different they could be considered the *opposite* of what we have seen so far. And the opposite to orderly growth implies a process that is either not growing or not orderly, or both.

In mathematical jargon, ergodic is an adjective meaning nonzero probability. Physics, true to its tradition of grafting mathematics onto real-life applications, has created the ergodic theorem which claims, in simple language, that situations which are theoretically possible will occur given enough time. By *possible* one can literally consider all imaginable situations that do not contradict fundamental laws. *Enough time* may signify that some situations may take practically forever before they come to life. Nevertheless, the theorem can be proven not only mathematically but also practically in everyday experience.

Consider fine glassware; its weakness is that it can break. Owners of fine glassware may take particular care, which can substantially extend the life of a cherished object, but one can be sure that sooner or later something that can break will break. It is only a question of time. This may sound like Murphy's popular law which states that whatever can go wrong will. But Murphy must have been in a cynical mood when he created it because his law lacks scientific rigor. The true ergodic theorem is amoral, inasmuch as it does not discriminate between good and bad, desirable and undesirable.

Obviously, the more unlikely the event, the longer it will take before it occurs. Imagine an enormous brightly colored moth flying in through an open window at night to land on your arm. Such an event may require waiting longer than your lifetime. In that sense it may never happen, but had you been able to sit there and wait in front of the window for thousands of years, you would eventually be guaranteed such a visit.

Ultimately, one might expect "miracles." As a physics student I had to calculate how long one has to wait in order to see a pencil fly! Since atoms in solids are vibrating in random directions, it is conceivable that there may be a time when all the atoms of one pencil direct themselves upward, in which case the pencil will lift off on its own. My calculation showed that had someone been watching a pencil without interruption from the day the universe was created, he most certainly would not have witnessed such a levitation. In fact, it is much worse; in order to have a fair chance for such a sighting, one would have to wait another period of that length, but this time with every second itself stretched to have the age of the universe! Yet the possibility is there, and it is only a question of time.

Apart from such academic fantasy games, the ergodic theorem can have more real implications in frequently encountered human behavior. It can be molded into a social law which says that if one *can* do something, one eventually *will* do it. For this to happen there will be a desire growing with time, pushing the person toward the act. In other words, capability breeds desire: *If I can, I want.*

Applications of this law in society abound. The famous answer to why climb a mountain, "Because it is there," is not quite complete. The correct answer is, "Because it is there and I can." After all, the mountain has been there as long as mankind. The same law applies to couples who are asked why they had children. Answers range from "We did not want to miss the experience" to "It just happened." In fact, according to this law, the answer is fundamentally the same in all cases: "Because we could." But people who do not recognize that their ability becomes a cause more often explain their actions in terms of conventional social behavior.

Another application of the ergodic theorem is a fact that is common knowledge among business managers: People at their jobs tend to do what they can rather than what they were hired for. It may sound chaotic, but some companies have been successful by allowing their employees freedom to shape their own jobs. Employees are happier that way. The law says that if they can do something, they will eventually do it, which means there will be a force—a desire—driving them in that direction. Blocking the realization of that desire can result in frustration, which may or may not be openly expressed.

What one can do changes with age, however, and so does the desire to do it. I was the supervisor of a man who smoked two and a half packs of cigarettes a day. Being an exsmoker myself, I suggested he quit. "I am still young," was the reply. It becomes easier to stop smoking as one gets older. This happens not because age enhances willpower but rather because the harmful physical effects of tobacco are tolerated less by a weakened organism. In other words, older people are less able to smoke. At the same time they do more of what they can do.

We are constantly faced with a multitude of things we can do. Our desire to do them is usually inversely proportional to the difficulty involved. Where does this argument lead?

· · ·

Ever since he was a little boy, Tim Rosehooded was drawn to mathematics. As a young man he became fascinated with the ergodic theorem and decided to give a deserving name to the law behind it: *Dum Possum Volo* (as long as I can, I want). Tim was also a seeker after knowledge across disciplines, philosophical schools of thought, and spiritual experiences. He searched beyond science, in metaphysics, esoterism, and the miraculous. When he found a group of people who had been working on fragments of what he thought was *real* knowledge, he began associating with them and diverted some of his time and energy toward the development of his inner self. One activity of this group was to spend an occasional Sunday doing spiritual work under the guidance of a master.

It was on such a Sunday morning that Tim was confronted with a conflict. It was a bright, clear day in the middle of the winter with lots of fresh snow on the surrounding mountains. Tim looked out his window. Ideal ski conditions, he thought. There were few of these days during the season. He liked skiing. He had just bought ski equipment that incorporated the latest technology. He paused for a minute in front of the window. He felt an urge to go skiing, but at the same time he felt bound by his decision to devote this day to spiritual work. At that moment he heard the words *Dum Possum*

Volo ringing in his head, as happened to him whenever he came across showcase applications of the ergodic theorem.

The arguments appeared spontaneously. There wouldn't be many days like this. He wouldn't be young and capable of skiing forever. He should take advantage of it now while it was possible. He could always go back to spiritual work when he would be old and unable to do the things he could now do.

Tim did not hesitate much longer. He went skiing that Sunday. He waited in lines, paid for tickets, got onto chair lifts, admired the breathtaking view, and took pictures. He felt like a tourist.

• • •

What Makes a Tourist?

The reason Tim felt like a tourist was that tourism stems from the if-I-can-I-want law; people visit a place simply because they are able to do so. Millions cross the oceans every year to visit even the remotest corners of the planet. As the number of those who can afford it goes up due to lower ticket fares and higher standards of living, the number of those who travel increases.

Christopher Columbus also crossed an ocean in 1492. He did it because he could, but in addition, and without being conscious of it, he was conforming to a necessity dictated by Europe's deeply rooted need to explore whatever lay westward from its known boundaries. When he sailed he was unaware of his contribution to the 150-year-long exploration process started 50 years earlier with several unsuccessful attempts. As we saw in Chapter Two, the overall learning process in Europe reached completion toward the end of the sixteenth century, and because Columbus participated in that process, he was an explorer, not a tourist—a distinction I will try to make clear.

Exploring is a process that follows a learning curve. It has a rate that goes through a life cycle. Making discoveries is characterized by an element of competition. It is *being first* that counts, just as with the breaking of a record. Once discovered, the same location can no longer serve as an object for exploration. As the opportunities for exploration become exhausted, competition for them leads to saturation.

In contrast, there is no competition and practically no saturation in tourism. The same places are visited over and over again. The rate here grows quickly to reach the limits of transport capacity, accommodations, and so forth. But from then on, fluctuations in the numbers of visitors are purely statistical. There are no learning curves behind being a tourist; it is governed solely by the ergodic theorem. The ability to be a tourist is the most profound reason for becoming one.

THE NUTRITIONAL VALUE OF NEW IMPRESSIONS

Explorers and tourists share a passion: a thirst for new impressions. At first glance the search for new impressions may appear to be a part of recreation. A change of scenery is usually agreeable and makes one feel refreshed. Peter Ouspensky argues, however, that new impressions have the highest nutritional value and can be considered an indispensable ingredient to life.[1] He classifies fundamental needs according to how much we depend on them. He considers ordinary food the least significant need since one can survive up to a month without it. Water is a more fundamental need because without it we can live only for a week. A still more essential need is air (oxygen), without which we die in a few minutes. Ultimately, the most refined and most essential "food" of all is impressions composed of a variety of external stimuli. Ouspensky claims that one may die in a matter of seconds if deprived completely of all sensory impressions.

It is difficult to verify practically the validity of such a hypothesis. There is some evidence that points in that direction. In sensory deprivation experiments, for example, subjects wear gloves and are left to float on water in a dark, soundproof cell. In a matter of hours they enter a temporary catatonic state, a condition similar to being unconscious, half-dead.

Further proof may lie in the fact that repeated identical stimulation (lack of variety) deadens sensitivity. Pickpockets put this into practice when they find ways to stimulate the wallet region of their victims before they pull the wallet out. Soon after putting on perfume, you no longer smell it because you have become accustomed to it. The iris of the eye oscillates rapidly and continuously to avoid stimulating the same cells on the retina. Looking at a uniformly blue sky can produce a sensation of blindness. The same effect is obtained if you cut a Ping-Pong ball in half and cover your eyes so that all visual stimulus becomes undifferentiated.

Finally, it is common knowledge how easily young children become bored. In urgent need to be nourished and grow, children seek new impressions ravenously. The more possibilities a toy offers, the longer they will spend with it, but they soon will turn elsewhere for more new impressions. Older people, on the other hand, do not search quite so insatiably for new impressions. Many prefer more meditative activities and a state of mental tranquillity rather than continual sensory stimulation. For an older person, an impression may no longer be new. Nevertheless, the person must still be nourished to stay alive and in good health, so a need for some form of sensory stimulation and new impressions continues throughout life.

The search for new impressions is the active agent behind tourism but also behind movie-going and visits to museums and expositions. The nourishment from having an experience for the first time generates pleasure. Most of the pleasure from photographs, for example, comes when one first looks at them or years later when all is forgotten and it feels as if one is looking at them for the first time. It is for new impressions that people travel, sometimes to improve their health on their doctor's recommendation. Wise schools of thought, often from Eastern cultures, teach techniques for looking at familiar things in "new" ways. In both instances the only beneficiary of the new impressions is the individual onlooker.

In explorations, however, like those of Columbus, the new impressions were not only for the benefit of the explorer. The whole known world was watching him. (This, by the way, was literally the case when Neil A. Armstrong stepped onto the moon for the first time and the whole world shared some of his first impressions through television.) The whole world benefited from Columbus's explorations through learning about a new world. But when the ten-millionth tourist visits Notre Dame in Paris, it is strictly the individual who will benefit from new impressions, and through his camera the world cannot learn anything new. Such an individual is a tourist, while Columbus, an explorer, acted according to a destiny, the learning process through which Europe discovered the new World.

FROM TOURISM TO S-CURVES

Tourism can sport a sophisticated disguise. I knew a gifted pianist who would play Bach pieces I had never heard before. To my surprise, several of these pieces turned out to be not only compositions of his own but on-the-spot improvisations. He had studied Bach extensively and was able to imitate his music perfectly. Similarly, there are painters who can make excellent reproductions of great masterpieces. Such works have low value but not necessarily because they are artistically inferior to the originals, as the experts invariably claim. They simply *are not* the originals. Beethoven did not become great just because his music was good. Besides being good, his music was also different from what the world was familiar with. Value and recognition are associated with *newness*.

Seen as a whole, tourism is repetitious behavior; tourists are all doing the same thing, copying one another. An activity like copying does not draw on someone's creative potential. There is no danger of running out of ideas. One can go on copying forever by copying copies. Creating, however, ultimately exhausts someone's innovative potential, is subject to competition, and follows a natural-growth process that at the end flattens out.

Exploration can be related to creativity and "newness," while tourism is related to copying and "the same old thing." If we consider the whole of Western European society as a single organism over time, we may be able to see exploration as an act of tourism. The exploration of the Western Hemisphere (Figure 2.2), when seen at this level of abstraction, can be interpreted as one action of the "organism" Europe. The organism takes such an action because it can—adequate ships, means of navigation, and other conditions permit it at some point in time—and for reasons of "nourishment," not unlike the tourist at Notre Dame. Alternatively, one may want to think of an American tourist as an explorer, for example, following his or her destiny on America's learning curve about a particular European monument. A conceivable saturation level can be envisaged when all Americans have passed in front of that monument.

One can thus try to relate natural growth to tourism but only on an abstract conceptual level. The difference in time scales and relative sizes is such that while the learning curves of natural growth provide a good description for explorations, they do poorly with tourism. The number of tourists in front of a monument per year is smaller than the rate of population increase, so saturation will never be reached. The cumulative number of tourists, no matter what time scale, will not look like an S-curve but rather like a straight line with some inevitable fluctuations superimposed.

Statistical fluctuations—random in nature—can be found in both rates of tourism and natural-growth processes. For the former this randomness is purer than for the latter, where fluctuations have to "arrange themselves" so as to average on the overall trend of the S-curve. Individual events may appear to be random, but groups of events must obey the fundamental law governing the natural-growth process.

Yet there is another phenomenon that bridges the gap between the randomness that characterizes tourism and the S-curves that characterize natural-growth processes: the phenomenon of chaos.

FROM S-CURVES TO CHAOS

Oscillatory behavior is one of the mathematical solutions of the Volterra-Lotka equations for a predator-prey system.[2] Such a system can be described as follows, with the lynx as predator and the hare as prey. The hare population grows at a constant rate (exponentially) in the absence of lynxes, while the lynx population declines at a constant rate—through starvation—in the absence of hares. When they coexist, the number of hares consumed is proportional to the number of lynxes, while the growth in the lynx population is proportional to the number of hares. This is a verbal description of the Volterra-Lotka system of differential equations.

Without having to go into the mathematical solution, we can see that starting from a situation where both species are scarce, the population of hares grows rapidly by reproduction. Then the lynx population also increases until there are so many lynxes that the hare population declines. This in turn causes the lynx population to decline, and with both populations low, the cycle starts again.

The example used here is realistic. In the northern forests the populations of these two species have been observed to fluctuate with a cycle of ten years, and the pattern observed for the lynx population lags the pattern for hares.[3] Differences from the ideal case occur, however, because other species interfere making these oscillations less regular but also increasing the stability of the ecological system.

Oscillations can also be observed in the absence of a formal predator—for example, in a rat colony that is given a fixed daily food supply.[4] The larger, more aggressive rats may start hoarding food in order to attract females when food becomes scarce. The timid rats huddle together, do not reproduce, and eventually die. The overall population decreases, and at the end only aggressive rats survive. They then propagate rapidly to reach a condition of an overpopulation of rats "enriched" in aggressiveness. A new cycle of hoarding may start, and eventually, extinction by degenerative belligerence may result.

But one can find oscillations in populations of much less aggressive species. Sheep introduced in Tasmania in the early nineteenth century grew in numbers to fill their ecological niche of about 1.5 million by the 1840s.[5] During the following one hundred years their population oscillated erratically around that ceiling, with diminishing amplitude. These fluctuations reflect changes in the birth and death rates, which in turn reflect the economy, epidemics, climatic changes, and other phenomena of a chaotic nature.

In scientific terms chaos is the name given to a set of ongoing variations that never reproduce identically. It was first observed when mathematical functions were put in a discrete form. Since populations are made up of discrete entities, a continuous mathematical function offers only an approximate model for the real situation. Discretization is also dictated by the need to use computers, which treat information in bits and pieces rather than as continuous variables.

There are many ways to put natural growth in a discrete form. In their book *The Beauty of Fractals,*[6] H. O. Peitgen and P. H. Richter devote a large section to chaos. The section, which comprises three-fourths of their book, starts with Verhulst[7] dynamics and ends with a discrete Volterra-Lotka system. They produce chaos mathematically by discretizing the natural law of growth described in Chapter One. They cast Verhulst's law into a *difference* rather than a *differential* equation. As a consequence the solution becomes a sequence of small straight segments. The overall population rises similarly to the continuous case, but now it does not reach the ceiling smoothly. It overshoots, falls back, and goes through oscillations. For some parameter values these oscillations do not subside; they either continue with a regular—simple or complicated— pattern or simply break into random fluctuations, producing chaos. Figure 10.1 is taken from their book. It shows three possible ways to reach the ceiling. The simplest one is oscillations of decaying amplitude converging in a steady state. The second case produces regular oscillations that do not diminish with time. The last case shows chaos—oscillations of no regularity whatsoever.

Well before chaos studies, ecologists had become aware of some erratic behavior in species' populations. James Gleick argues that they simply did not want to admit the possibility that the oscillations would not eventually converge to a steady state level. The equilibrium was for them the important thing:

J. Maynard Smith, in the classic 1968 *Mathematical Ideas in Biology,* gave a standard sense of the possibilities: Populations often remain approximately

REACHING THE CEILING IN A DISCRETE SYSTEM MAY LEAD TO CHAOS

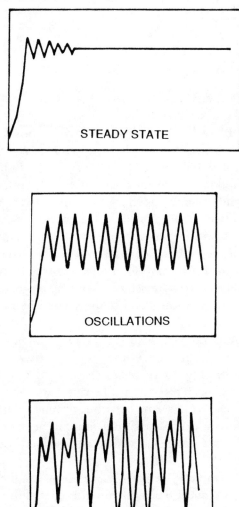

FIGURE 10.1. Three different solutions to the discrete form of the natural growth equation. The initial rise corresponds to the familiar S-curve of the continuous case. What ensues after reaching the ceiling may be stable, oscillatory, or chaotic.*

*Adapted from a graph by H. O. Peitgen and P. H. Richter in *The Beauty of Fractals* (Berlin and Heidelberg: Springer-Verlag, 1986). Reprinted by permission of the publisher.

constant or else fluctuate "with a rather regular periodicity" around a presumed equilibrium point. It wasn't that he was so naive as to imagine real populations could never behave erratically. He simply assumed that erratic behavior had nothing to do with the sort of models he was describing.[8]

Outside biology there have been many observations of data deviating from the idealized growth pattern once the 90 percent level of the ceiling has been reached. Elliott Montroll gives examples from the United States mining industry in which the annual production of copper and zinc are known to fluctuate widely over the years.[9] These instabilities can be interpreted as a random search for the equilibrium position. The *system* explores upward and downward while "hunting" for the ceiling value. If the optimal level is found, an oscillation may persist as a telltale signal of the regulating mechanism. If such a level cannot be found, erratic fluctuations may ensue, resembling chaos.

An example from the automobile industry is shown in Figure 10.2. The annual number of new registrations in Japan grows following an S-curve quite closely until the knee of the second bend, where it starts fluctuating.

These fluctuations do not necessarily have an impact on the total number of automobiles in use. As mentioned earlier, the car niche in Japan, like that of most Western countries, is saturated. New car registrations mainly represent replacements. During hard times people tend to hold on to their cars. New car registrations dropped significantly in the United States during World War II, but the number of cars in use—the fundamental need—did not deviate from the S-shaped course it had been following.

Despite the fluctuations in new car registrations, the lifetime of a car is, under normal circumstances, rather stable and rigidly links the appearance of replacements to the original S-curve along which the automobile niche was filled up in the first place. This fact provides a means for calculating car demand *independently* of economic considerations. With the lifetime for cars as a constant, one can compute a new-registrations curve by replacing cars as they die in the original population curve. Such a simulation was carried out for the case of Japan; it reproduced successfully the data pattern of Figure 10.2 but not the chaotic fluctuations at the ceiling.[10]

Instabilities at the top of the curve should be expected only when we are dealing with *rates* of growth. Accumulation of growth is different. A

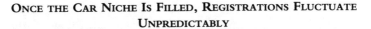

ONCE THE CAR NICHE IS FILLED, REGISTRATIONS FLUCTUATE
UNPREDICTABLY

Annual registrations
in millions

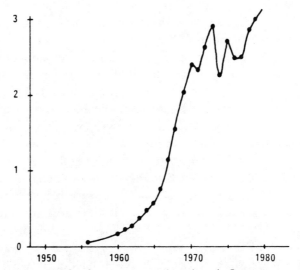

FIGURE 10.2 Pure data (no fit) for new car registrations in Japan as reported by Cesare Marchetti.*

cumulative quantity either grows or stays flat. The number of known elements—the works of an artist, the number of hamburgers sold up to a given time, or the height of a child—is a quantity that cannot display a downward dip.

A characteristic chaotic pattern also emerges in the annual rate of plywood sales. In an article published by Henry Montrey and James Utterback there is a graph showing the evolution of plywood sales in the United States (Appendix C, Figure 10.1). First commercialized in the 1930s, plywood started filling a niche in the American market. From 1970 onward a pattern of significant (plus or minus 20 percent) instabilities appeared, which Montrey and Utterback tried to explain one by one in terms of socioeconomic arguments. Given a certain pattern one can always correlate other phenomena to it. This type of pattern, however, could have been predicted a priori by chaos formulations.

*From Cesare Marchetti, "The Automobile in a System Context: The Past 80 Years and the Next 20 Years," *Technological Forecasting and Social Change*, vol. 23 (1983):3–23. Copyright 1983 by Elsevier Science Publishing Co., Inc. Reprinted by permission of the publisher.

In addition to observations of chaos when approaching the ceiling value, deviations between data and S-curves are also seen in the early phases of the growth pattern. In an anthropomorphic interpretation we attributed them earlier to "infant mortality" or "catching up." In growth in competition, the times before reaching 10 percent of the maximum can be seen as a trial period during which survival is at stake and major readjustments may be undertaken. In industry, for example, products are often repositioned by price changes shortly after launching. Other deviations observed in the early phases of growth are due to the release of pent-up energy following delays usually attributed to "technical" reasons. Many of the life curves of the creative careers shown in Chapter Four feature such an early accelerated growth, which was interpreted as an attempt to make up for lost time.

This is a behavioral explanation of the irregular patterns often encountered at the extremities of S-curves. Chaotic behavior dictated by the discrete nature of populations can account for half of these irregularities, the erratic behavior of the ceiling. It does not account for the early deviations. Alain Debecker and I obtained a mathematical understanding of this phenomenon by linking S-curves to chaos in a way different from the classical approach described above. It happened while we were looking into the chaotic behavior in the annual production of coal in the United States as shown in Figure 10.3.

Production of bituminous coal increased naturally during the better part of a hundred years. After 1920 production reached a plateau and the familiar erratic oscillations appeared, but the last part of the graph revealed a "disturbing" systematic increase since 1950. Is it an exceptionally large chaotic fluctuation, or is there a new niche opening up?

Before and After Chaos

Chaos studies, as they are carried out today, focus on what happens *after* a niche is filled. The initial rise is largely ignored by chaos scientists, who are interested only in phenomena with erratic fluctuations. Many real cases, however, such as the production of coal shown above, display multiple periods of growth. New markets open new niches, and the historical picture of coal production looks like a succession of S-curves. Between two successive curves there may be a period of instability characterized by chaotic fluctuations.

Debecker and I tried to reproduce mathematically the chaotic patterns seen in the coal production data. We knew that in order to observe chaos

TEMPORARY "CHAOS" IN U.S. COAL PRODUCTION

Billions of tons

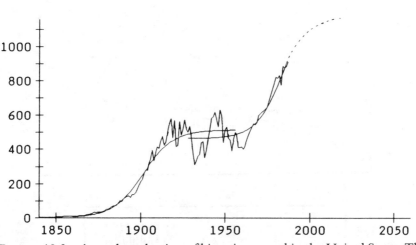

FIGURE 10.3   Annual production of bituminous coal in the United States. The two S-curves are fits to the data of the respective historical periods. The interim period shows large fluctuations of a chaotic nature.*

we had to pass to the discrete case, but instead of tampering with Verhulst's law, the differential equation, we put into a discrete mathematical form its solution only, the S-curve. By doing that we obtained instabilities at both ends of the growth process (see Figure 10.4).

In real-life situations we cannot see the early oscillations completely because negative values have no physical meaning. We may see, however, a precursor followed by an accelerated growth rate, an overshoot of the ceiling, and finally erratic fluctuations. These phenomena are interrelated. Precursors are not simple cases of "infant mortality." Their size and frequency can help predict the steepness of the catching-up trajectory, which in turn will reflect on the importance of the overshoot when the ceiling is approached.

The state depicted on the bottom graph is one of only *approximate* chaos because the erratic behavior is diminishing with time; the fluctuations will die out sooner or later. Also, our description is reversible; from late patterns one can in principle recover early ones, which would have been impossible with true chaos. Still, there is an advantage in our method

*The data come from the *Historical Statistics of the United States, Colonial Times to 1970*, vols. 1 and 2, Bureau of the Census, Washington, DC, 1976, and the *Statistical Abstract of the United States*, U.S. Department of Commerce, Bureau of the Census, 1986–91.

MAKING AN S-CURVE DISCRETE

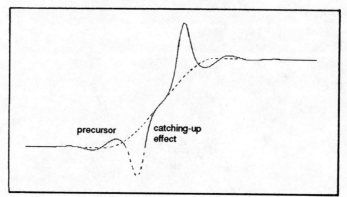

precursor   catching-up effect

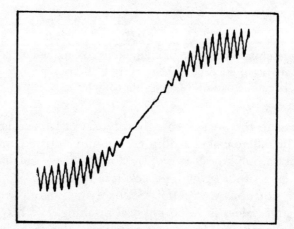

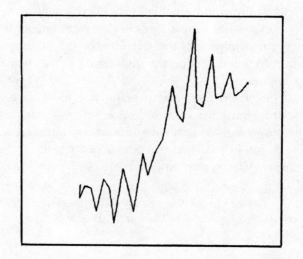

because we observe erratic behavior that precedes as well as follows the main growth period. Our approach provides a model for *all* the instabilities connected to a growth process.

In the light of Figure 10.4, the large oscillations in the production of bituminous coal may be seen as belonging to the ceiling of the first S-curve *or* to the beginning of the second one. In fact, if we had no knowledge of history before 1940, we would have characterized the early fluctuations of the more recent S-curve as "infant mortality" followed by an early "catching-up" effect.

A more typical example of growth where order alternates with chaos can be found in the graph depicting per-capita annual energy consumption worldwide that is presented in an article by J. Ausubel, A. Grubler, and N. Nakicenovic (Appendix C, Figure 10.2). The graph shows a niche-filling process from 1850 to 1910, followed by a chaotic state from 1910 to 1945, a second niche filling from 1940 to 1970, and a second chaotic state starting in 1970 and projected to continue through the year 2000. The authors, sufficiently aware of this alternation, continue to predict the end of the second chaotic state, the opening of a third niche, and a third chaotic state occurring during the second half of the twenty-first century.

The chapter on linking order to chaos is far from being closed. More investigation is needed both in case studies and in theory. The evidence presented here suggests that conditions of natural growth alternate with states of erratic fluctuations. A well-established S-curve will point to the time when chaotic oscillations can be expected—when the ceiling is being approached. In contrast, an entrenched chaos will reveal nothing about when the next growth phase may start. One has to locate it from other considerations particular to each case. The fifty-six-year economic cycle is one way to predict growth periods. Both coal production and

---

FIGURE 10.4   These are examples of the patterns we obtained by putting a natural-growth curve in a discrete mathematical form. The figure at the top demonstrates the so-called early catching-up effect and the possibility of precursor and overshoot. In the middle we see regular oscillations, and at the bottom, states resembling chaos. In all cases, similar behavior is observed early as well as late in the growth process.*

*Adapted from Theodore Modis and Alain Debecker, "Chaoslike States Can Be Expected Before and After Logistic Growth," *Technological Forecasting and Social Change,* vol. 41, no. 2 (1992).

energy consumption are seen to rise from 1884 to 1912 and again from 1940 to 1968, periods of rapid economic development worldwide.

Generally speaking, S-curves are important tools in modeling the stage of growth, while descriptions of chaos are more appropriate for the irregular fluctuations observed in the absence of growth. There are phenomena for which the initial rise in the growth pattern becomes quickly irrelevant—for example, building up a smoking habit. Every smoker starts from zero, but most people's concern is how much they smoke per day rather than the detailed path along which their habit was formed. The number of cigarettes smoked per day displays chaotic fluctuations. On the other hand, when it comes to growth in a child's height, the small fluctuations during adolescence, after the end of the process, are of no consequence whatsoever.

Thus, there are aspects of human behavior in particular that do not follow the S-shaped pattern characteristic of natural-growth processes. Human behavior such as tourism has its origin in the law that says if I can do something, I want to do it. The rate of tourism has no life cycle associated with it; its fluctuations are of a purely statistical nature. Statistical fluctuations are an inseparable feature of repeated measurements. One should also expect chaotic fluctuations when observing a natural-growth process, but only during the early and late low growth phases. During the phase of the steep rise, fluctuations play a secondary role; the limelight in this period is on the steady rise of the growth pattern, which, unlike the fluctuations, *is* predictable.

11

Forecasting Destiny

• • •

It was a usual get-together for the employees of the company's customer services department. The occupants of the small building had all gathered in the large conference room to listen to announcements of new policies, promotions, office moves, welcomes, and farewells. This time there was a lot of talk about desks and office equipment, culminating as follows:

"We all know how painful it has been to get our coffee from a machine that requires exact change. How many times has each of us run around asking colleagues for coins? Well, you can consider this activity part of the past. Next week we are installing new coffee machines that accept any combination of coins and provide change. This should improve employee productivity and satisfaction."

In spite of its banality, the statement resonated in me because I had indeed been frustrated more than once by not having the correct change for my morning cup of coffee. The announced improvement had been long overdue. The following week the new machines were installed and for several days employees made jokes about mechanical intelligence and technological comforts.

It was several weeks later that I found myself in front of the machine without *enough* change for my mid-morning coffee. I looked in my desk drawers for any odd change—in earlier times there would have been "emergency" coins lying around—but with our new machines, there was no longer any reason to hoard change. My colleagues were in a meeting, and I did not want to interrupt them with my petty problem. I walked over to the secretary's desk; she was not there. Instinctively, I looked in the box where she used to keep piles of change for coffee; it was empty.

I ran out of the building to make change at a nearby restaurant, wondering why those coffee machines couldn't accept paper bills. Then I realized that the old machine had not been so inconvenient after all. It had been a nuisance to have to plan for a supply of correct change, but since that was the reality, we all made provisions for it in some way or another. It took some forethought, but one could not really say it was a stressful situation (being deprived of a cup of coffee never killed anybody). With the installation of the new machines, however, we relaxed more. Perhaps I had relaxed a little too much. I decided to make sure that in the future there is always *some* change in my desk so as to avoid situations like this.

At the same time I felt there was a certain quota of inconvenience associated with getting a cup of coffee. If the amount of inconvenience rose above the tolerance threshold, actions were taken to improve the situation. On the other hand, if obtaining a cup of coffee became technically too easy, we would relax more and more until problems crept in to create the "necessary" amount of inconvenience. Imagine a situation where coffee is free and continuously available. A busy person, trying to be efficient and exploit the fact that there is no longer reason to "worry" about coffee, may simply end up not planning enough time to drink it! Perhaps people are not happy with too little inconvenience.

· · ·

Many people claim that they must spend all the money they make in order to cover their needs. Yet this becomes the case again soon after they get a sizable salary increase. The usual rationalization is that their needs also grow so that they still spend their entire income. The fact, however, that they always spend just what they earn, no more, no less, points to an equilibrium in which the needs grow only to the extent that there is more money. The same may be true of work, which expands to fill the time available. Some people, for example, continuously find things that need to be done around the house or apartment, independently of whether they went to work that day or spent the whole day at home.

The concept of an invariant was introduced in the beginning of this book to describe an equilibrium of opposing forces, or a tolerance level. It can also be seen as the ceiling of a niche that has been filled to capacity. If the growth is natural, the process should reach completion because niches in nature do not normally remain *partially* filled. One has to be open-minded about what may constitute a niche. In the office situation described above, there was a "niche" of inconvenience in getting a cup of coffee. This niche remained intact. When the company took action to lower its ceiling, employees became more negligent so that the level of inconvenience rose again, as if the old level was not only well tolerated but in a way "desirable." Still, the action taken by the company eliminated one activity: stockpiling exact change.

Equilibrium levels, such as those represented by the level of a niche filled to capacity, are regulated by some mechanism. Therefore, at close examination they always reveal oscillations. For example, the very first invariant mentioned in Chapter One, the annual number of deaths due to automobile accidents, oscillates around the number twenty-four per one hundred thousand population in most Western countries. When the number rises to twenty-eight, social mechanisms are activated to bring it back down; if it drops low enough, attention drifts to other concerns, allowing it to rise again.

The "naturalness" of a process can thus be evidenced in two ways: by an unambiguous S-shaped pattern of growth, but also by persisting oscillations around a constant level. In fact, the latter is a partial manifestation of the former since it is what happens at the beginning and at the end of the natural-growth S-curve. One way or another, at any given moment of the process, there is a clear indication about what projections into the future can be made.

All along the trajectory one can find superimposed fluctuations resembling chaotic phenomena, themselves claiming a right to naturalness. One could debate endlessly about which one is more natural, the S-shaped growth pattern or the random fluctuations. But from a practical point of view it must be pointed out that chaos studies have not yet demonstrated improvements in forecasts. Then, too, deviations from the natural-growth pattern may result from wars, governmental actions, and fads. But such deviations can be characterized as unnatural; they are relatively short-lived, and the growth process soon regains its previous course. It might be revealing to examine some cases of these unnatural deviations, the individuals who are responsible for them, and the consequences of their actions.[1]

DECISION NON-MAKERS

A country's government is a typical decision-making body that sets courses of action for an entire nation. Enforcement of a new policy may seem desirable and feasible in the minds of the country's leaders, but when policy decisions are made with no respect for the natural processes already established, the results can be negative and provoke embarrassment if not violent opposition. I will give two such examples, both of them concerning the important subject of planning for primary energy sources.[2]

The first example takes us to the United Kingdom, one of the world's major coal producers. In the substitution of primary energy sources presented in Chapter Seven, we saw that the relative importance of coal has been declining in favor of oil (and to a lesser extent natural gas) since early in this century. This substitution is natural and valid in general even if the amount of energy obtained from coal today is far from negligible. Coal's share of the total primary energy market has been decreasing worldwide for the last fifty years. Coal production in the United Kingdom started declining in 1950 and followed a usual "phasing out" course (see Figure 11.1). The projection suggests that production should drop to less than 20 million tons a year by the end of the century. For the government of the United Kingdom, however, such a vision may be completely unacceptable.

In 1975, as a mechanism to absorb oil shocks, the government halted the decline in coal mining by a legislative act that fixed production at 125 million tons a year. The act caused a clear deviation from the declining course of coal production, which lasted nine years. At that time miners staged the longest strike ever, bringing coal production down. The fact that the level of production dropped to what it should have been had it followed the pre-decree trajectory makes one wonder whether renewed high-production levels portend another major action by miners.

The second example concerns oil imports in the United States. Figure 11.2 shows the relationship between imports and total domestic production.[3] Before the oil crisis of 1969 there was a law that limited imports to a fixed percentage of the total amount of oil available. As a consequence we see a flat percentage up to 1969.

THE DECLINE OF COAL PRODUCTION IN THE UNITED KINGDOM

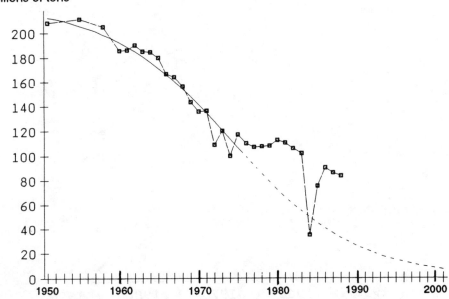

FIGURE 11.1   Annual coal production in the United Kingdom, with an S-curve fitted only to the data of the historical window 1950–75. The miners' action of 1984 went in the direction of restoring the natural decline halted by legislation in 1975.*

The fact that the quota was always met indicates a certain pressure on the constraining law. It was probably due to this pressure that the quota was lifted in 1970. At the same time an ambitious energy project was put in place to decrease the country's dependence on oil imports. What followed was a sharp rise in the percentage of imported oil, which frustrated and embarrassed the architects of the energy project. The rise continued for ten years, reached the value of 60 percent, and then began to decline, evidence of having overshot the ceiling of the niche. The large amplitude of the ensuing oscillation may have been a consequence of the prolonged repression of imports.

The country's leaders failed in their attempt to enforce what they thought was a "good" level for oil imports. As demand increased and could not be met cost effectively by domestic production, the quota law had to be relaxed, and the energy independence project proved ineffec-

*The data on U.K. coal production come from *The Annual Abstract of Statistics,* a publication of the U.K. Government Statistical Service.

THE DIFFICULTY IN IMPOSING GOVERNMENT'S WILL

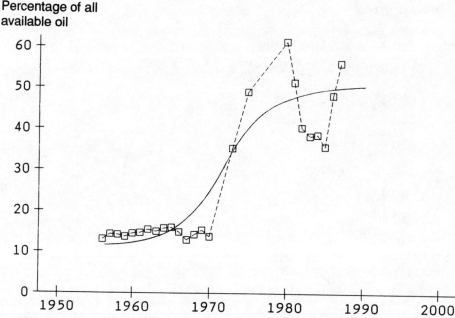

FIGURE 11.2   United States oil imports as a percentage of all available oil—imports plus domestic production. A quota on imports had to be relaxed in 1970, and the ambitious energy independence project launched at the same time proved ineffective. The superimposed S-curve is not a fit to the data but simply an idealized natural growth scenario.*

tive. The leadership's only achievement was short-term deviations from what should have been a natural course.

In discussing this example of the ultimate futility of attempting to alter the direction of a natural course, Marchetti presents a satirical image of the world's most powerful man, the American President, who is "like Napoleon in Russia, sitting on a white horse pointing east while the army is going west; and that's not the best image for decisional power."

War has been held responsible at times for starting new trends and establishing new processes; however, it provides conditions that can hardly be considered natural. We saw earlier that during World War II natural rubber became in short supply and efforts intensified to fabricate rubber. The feat was accomplished, and natural rubber began to be

*The data come from the *Statistical Abstract of the United States,* U.S. Department of Commerce, Bureau of the Census, and from *Historical Statistics of the United States, Colonial Times to 1970,* Bureau of the Census, Washington, D.C.

replaced rapidly by synthetic. But as soon as the war was over, production of synthetic rubber dropped, despite the acquired know-how and the ability to produce it in large quantities. Natural rubber appeared again in the market, and the substitution process continued at a slower, more "natural" rate. The ratio of synthetic to natural rubber reached the World War II levels again only twenty years later, in the late 1960s.

Deviations from a natural path may also occur in peacetime and involve more than one government. The worldwide diffusion of nuclear energy during the 1970s followed a rate that could be considered excessive, compared to previous energy sources. In Figure 7.6 we saw the market share of nuclear energy grow along a trajectory much steeper than the earlier ones of oil, natural gas, and coal. This abnormally rapid growth may have been what triggered the vehement reaction of environmentalists, who succeeded in slowing it down. They are far from having eliminated it, however. The projections in Figure 7.6 show that nuclear energy has a niche for itself, has already gone beyond the "infant-mortality" stage, and will grow to reach maturity sometime in the twenty-first century.

The nuclear energy controversy presents other intricate facets. In a lengthy study by Marchetti entitled "On Society and Nuclear Energy,"[4] we enter a kind of twilight zone: decisions with no one responsible for them, effects that look like causes, a wild card for the role of a decision maker, and accidents that will happen because they *can* happen given time.

The construction of nuclear reactors in the United States has come in waves. The first wave consisted of seventy-five nuclear power stations, with their coming into operation centered around 1974. The second wave involved forty-five sites, and they have not yet all come into operation. More construction waves are likely to follow if nuclear energy becomes as important in the future as coal and oil have been in the past. This clustering into spurts of nuclear energy activity makes it possible to focus on the web of relationships among the number of power plants, nuclear accidents, the press, and ordinary citizens.

Besides the malfunction of equipment, nuclear accidents are also caused by purely human errors. In these cases the psychological, emotional, and mental state of the operators can play an important role. Reactor operators are men and women who happen to work at nuclear sites but are otherwise ordinary people. They have families, go home after work, read the papers, and watch television. They are part of the public. Thus, when we consider nuclear accidents we should also look at the media. The

media is involved in a feedback loop with the public. On one hand the
media helps define the shape of public opinion, but on the other hand it
reflects what the public likes and is waiting for. This feedback mechanism
makes the media behave as a resonating cavity for public opinion and
concerns. In feedback systems, causality can be seen *both* ways: Accidents
make stories, but stories may cause accidents.

There is evidence to support this hypothesis. In Figure 11.3 we see two
growth curves: one for the appearance of major nuclear plant accidents,
and the other for press coverage of the subject of nuclear energy in
general. The stories regarding both the pros and cons of nuclear energy

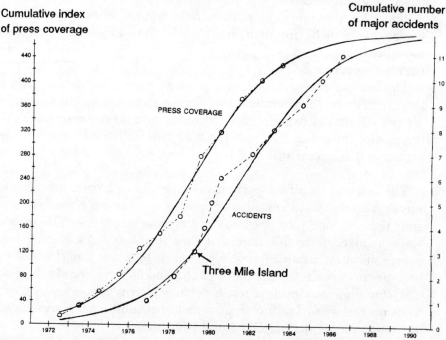

FIGURE 11.3   Two sets of data and respective fits. The one on the left is an
annual index of press coverage, calculated on the basis of space devoted to the
subject of nuclear energy among American periodicals. The units shown on the
left vertical axis are arbitrary but normalized to 100 for the year 1979 during
which publicity about nuclear energy received maximum attention. The data on
the right concern the major nuclear accidents over this historical period.*

*Reported by Cesare Marchetti in "On Society and Nuclear Energy," report EUR
12675 EN, 1990, Commission of the European Communities, Luxemburg.

simply reflect the attention paid to the subject. The graphs concern the United States, a country that claims both the largest number of reactors in operation and the most years of experience; at the same time it is a country in which the press is particularly dynamic and influential. The cumulative accidents data are for nuclear plants built during the first wave of construction (circa 1974). The sites in the second wave have not been in operation long enough to produce major accidents. The press coverage is quantified in terms of an index representing the percentage of periodical space devoted to information relating to nuclear plants.[5] The cumulative index is plotted in the figure. The curves fitted on the data provide a fair description for both accidents data and the coverage index, indicating that we are indeed dealing with phenomena that have a well-defined beginning and an end, all relating to the first wave of reactor construction.

The press coverage curve is parallel to, but comes two years *before,* the accidents curve. Opinion intensity seems to have preceded accident probability by a couple of years. This may not prove that it was the press coverage which caused the accidents (chronological succession does not necessarily mean causality), but it disproves the opposite: that it was the wave of accidents which stimulated the press coverage.

It could be that a kind of *cultural epidemic* renders nuclear reactor operators more accident-prone. David P. Phillips has provided evidence in support of what he calls the "Echo Effect," claiming that following the publication of news of the suicide or homicide of a popular personality, car accidents—single cars in particular—increase significantly with respect to the usual average. They show a large peak three days after the report and a smaller one eight days later. Similarly, there is a three-day strong echo effect (and a weaker seven-day one) in airplane accidents—many of them corporate planes (Appendix C, Figure 11.1). Most of these accidents are suicides in disguise.

There are some interesting coincidences concerning nuclear plant accidents.[6] The film *The China Syndrome* describes events similar to those that occurred during the Three Mile Island accident; it was released just a few weeks earlier. The accident itself occurred at the peak of the twelve-year-long press coverage curve (the middle of the S-curve). This accident, which appeared as the worst ever in the United States, received the most publicity and produced the strongest emotional response. Could all this publicity have contributed to putting reactor operators around the country in a precarious psychological condition? The next three accidents came in such close succession that they deviated from the natural curve (see Figure 11.3). In the late 1980s the sinister curve has flattened. Unlike the early 1980s, when we saw five major incidents in three years, we now expect only one in five years.

To recapitulate, we have seen in this section a variety of situations, ranging from miners' strikes and energy programs to the necessities of war and nuclear accidents, in all of which events seemed to follow a different course from the one set by the decision makers, be they in the government, in the military, in technology, or in science. In the usual perception, leaders are faced with multiple alternatives and have to choose among them. If indeed they can choose, it means that there are many paths along which events may proceed. But how many paths can there be? Which ones will merely produce a short-term deviation from what will later be recognized as a natural trend? Some time ago J. Weingart and Nebojsa Nakicenovic made a compilation of predictions for solar energy and total energy demands according to different models.[7] They produced trajectories for seven different scenarios covering a wide range of futures. The scenarios diverge, and by the year 2200 there is a factor of five between the most pessimistic and the most optimistic one. How can there be *many* futures? There has been only *one* past.

Theoretical physicists like to build abstract models. In one of them the idea of parallel worlds says that a multitude of alternate ways to continue today's world springs into existence every second. All these worlds evolve and multiply in infinite directions every second. Such a science fiction view of reality is entertaining. Furthermore, it endorses the often-cited morality of the unlimited number of choices the future has in store for us. But we are confined to one world, and there is simply no way to communicate with or produce evidence for the existence of the other ones.

In our world the evolution of energy substitutions generally followed well-defined natural trajectories. Deviations from these trajectories should be interpreted as "hunting" for the path. But hunting implies a hunter. The future scenarios of the different models may all be possible. They are the mutants. They exist because they can. (Presumably these models are not based on impossibilities.) Our future will be unique, however, and will be based on the mutation best fit for survival. The path is built through a trial-and-error approach. If a choice is good, it is kept, developed, and diffused; if it is not, it is rejected in favor of a chance to try another choice. But the one who decides is not a government, a leading scientist, or another decision maker. It is the system itself: the complete set of interrelated physical quantities that have evolved via natural-growth curves.

A System's Predictability

The energy system has worked well so far. Coal was there when we needed it, and oil is available within a few miles of wherever you happen to be. Oil crises have come and gone without interfering significantly with everyday life. When domestic production was not enough, imports increased "naturally," and whenever the oil-producing countries tried to devise mechanisms to raise prices (including waging war), they achieved at most short-term price fluctuations. The Iraqi war in 1991 provoked an ominous rise in the price of oil, but it did not last more than a few weeks. An energy price hike like the one of 1981, lasting many months or years, is a periodic phenomenon and, according to the fifty-six-year economic cycle, is not expected to happen again before well into the twenty-first century.

The energy system worked well even when energy prices skyrocketed in 1865, 1920, and 1981. On these occasions there was some economic impact, but it interfered minimally with the functioning of society. The trends in satisfying energy needs show no noticeable deviations at the time of the price hikes (see Figure 8.3); the trajectories are stable and therefore predictable.

We often hear that the future is not predictable. This statement if unqualified contradicts the fact that no one packs an umbrella for a summer vacation in Greece, and even more so the fact that it was possible to predict the time of lunar landings within seconds. The correct statement is, "The future is not predictable with infinite precision," and the reason is that prediction involves calculation, which entails handling information, which requires energy. To make a prediction with more and more precision requires more and more energy, and becomes impossible beyond a certain level. This is not a reference to Werner Heisenberg's celebrated uncertainty principle which postulates that it is impossible to specify an object's location and momentum—or energy and time—with unlimited precision. Trying to predict global energy needs by tracking down the energy consumption of each individual is similar to trying to predict gas pressure in a vessel by tracking individual molecules. Both processes involve calculations that run into energy problems despite today's powerful computers. Computers are after all inefficient in their use of energy.[8]

Wasteful Computers

The central processing unit (CPU) of a computer—its brain—uses electricity to handle information. The amount of electricity it uses may be negligible compared to other machines or even to its peripherals: printers, terminals, and so forth. The amount of work it does is also very small. It manipulates bits of information. Information can be related to energy theoretically through the concept of negative entropy, the amount of order. As the order decreases in a system, the amount of meaningful information in it diminishes, and so does its energy. The implication in simple terms is that knowledge is paid for with energy. The smallest bit of information is a yes/no knowledge, the on/off state of a transistor. In units of energy (ergs) it has an energy content equal to 2×10^{-14}.

Efficiency in thermodynamic terms is defined as the ratio of output (useful energy) divided by the fuel (energy input). The efficiency of a computer's CPU can then be defined as the ratio of the energy equivalent of one bit of information divided by the amount of electricity used to change the state of one transistor. This ratio turns out to be in the range of 10^{-11} to 10^{-10}. It means that, relatively speaking, computers generate enormous amounts of heat for their calculations. This can become a formidable obstacle for applications involving enormous numbers of calculations.

Miniaturization makes computers more efficient because the higher the density of components on an integrated circuit, the lower the consumption of electricity. Economists would justify miniaturization through the money it saves and argue that the price of energy should have an impact on the evolution of efficiency. One can prove them wrong by demonstrating that the evolution of efficiency has deeper roots and long-term cultural links, and is not affected by the availability of energy or changes in its price.

The evolution of efficiency over time documented with data points for three selected technologies yielded three distinct straight-line segments (plotted in the nonlinear scale introduced in Chapter Six). One was for prime movers (engines), with its beginning around 1700. Another one, the technology for making light, started in the days of the candle. The third one was the technology for producing ammonia in the twentieth century. Today all three are still below the 50 percent level (Appendix C, Figure 11.2).

Becoming more efficient is a learning process; consequently the evolution of efficiency is expected to follow an S-shaped natural-growth curve. The fact that the data points fell on straight lines shows the natural-growth character of the processes. The data adhered to the natural-growth paths even though energy prices surged manyfold on at least four different occasions during this period. There is an internal regulatory mechanism in the evolution of technology that points to a deep-seated and stable organization, just as with the energy system. In fact, technology and energy can be thought of as subsystems of the greater *human system*.

Being 100 percent efficient is a God-like quality. In that respect humans still have a long way to go. In spite of the specific gains in efficiency over the last three centuries, if we take all energy uses together in the developed countries, we find that the overall efficiency for energy use today is about 5 percent. It means that to produce a certain amount of useful work we must consume twenty times this amount in energy. This result makes humans look better than computers but still shamefully wasteful, and raises worries in view of the magnitude of the useful work achieved daily.

To appreciate this more, consider the following question: How much have Americans increased their per-capita energy consumption in the last one hundred years? The answer of well-informed people in general is between a factor of ten and a factor of fifty. The real case is only a factor of two! How can that be? Because in the meantime all processes involved in obtaining *useful* work have improved their efficiency. Today's average efficiency of 5 percent may well be five times higher than that of one hundred years ago, and compensates for the increase in demand for primary energy.

Increasing efficiency is probably underestimated when dealing with energy problems. The total primary energy consumption in the United States is expected to increase by a factor of two to two and a half between now and the mid-twenty-first century. *Useful* energy, however, will most likely jump by a factor of twenty, considering that we are only at the beginning of the efficiency growth process, which still resembles an exponential increase.

Energy planners should take into account the regular and forecastable evolution of our efficiency. In producing their models they would make a mistake if they underestimated the built-in wisdom of the system. The world primary energy substitution picture shown earlier in Figure 7.6 suggests that a "natural" phaseout of the old primary energy sources and a phasing in of nuclear energy, with a possible new source around the year 2020, could provide a smooth transition to the twenty-first century.

The problems do not seem to lie with physical resources but rather with international cooperation, an area where decision makers could be more effective.

Most energy forecasters focus today on the technological side of the energy picture. They are trying to find in technology all causes, effects, and solutions. Marchetti strove to redress the balance a little by over-stressing the importance of the frame. He summarized his advice in a warning: "Don't forget the system, the system will not forget you."

The French have a delightful fable written by La Fontaine about a busy fly that buzzes around the horses pulling a heavy carriage up a hill. Once at the top, the fly feels exhausted but self-satisfied with his good work. This fable seems to apply to much of the behavior of decision makers. One wonders how many "burdensome decisions" had to be made during the last 150 years to keep the market shares of primary energy sources so close to the straight lines of Figure 7.6.[9]

LITTLE CHOICE

Predictability of a system's behavior implies a certain amount of prede-termination, which is a taboo in Western society, particularly among forceful, strong-willed individuals of the kind typically encountered in a dynamic company's marketing department. These people like to plan out the future in long, heated discussions in which forecasts are heavily influenced by the strongest personality present. Marketers are people with enthusiasm; their scenarios of the future are often biased toward flattering their egos with notions of possessing the power of free will. My experience is that marketers' long-term forecasts also tend to be system-atically biased in favor of their new products and against the old ones. In general they fail to pay adequate attention to the *natural* progression of well-established processes such as substitutions, phasing-ins, and phasing-outs.

My first frustrating experience with marketers was when I confronted them with notions stemming from observations of stability in the auto-motive industry; for example, that innovative gadgets and price wars have little impact on car buyers, as can be evidenced by the very smooth growth of car populations. (The number of cars in use in the United States did not decline during World War II when car sales dropped to zero.) Such growth curves are shaped by fundamental needs and are little influenced by the shape of new bumpers.

The marketers were not happy with that notion. "You are telling us that no matter what we do, things will go their way" was their angry objection.

No, there are things to be done and those who market products do them most of the time, but they believe they are shaping the future, while in reality they are only compensating for changes in the environment in order to maintain the established course. They are reacting to change in the same way a biological system would in nature. Whenever there is a danger of falling behind the established path, they bring out good ideas from manufacturing, product design, and advertising. These ideas may have been dormant for some time; they are called upon as needed. This is how it happens in biological systems. Mutants are kept in a recessive state until a change in the environment brings them into the foreground.

From this point of view, innovation and promotion are no longer seen as aggressive operations but as defensive tactics. They can be thought of as ways of maintaining the equilibrium, just as sweating and shivering help regulate the temperature of the body. Innovation and promotion do not create new markets. They are part of the competitive struggle, which becomes more and more important as growth processes saturate, and the only way to maintain "healthy" growth is by taking the food from a competitor's mouth.

Unlike biological organisms, which can be healthy without growing, industries are healthiest while expanding rapidly. During this phase the workforce has a vested interest in increasing productivity, which will justify higher salaries. On the other hand, during the low-growth phase— toward saturation—productivity slows down because the workforce "feels" that increasing it will lead to a reduction in jobs and, consequently, unemployment.

But coming back to the question of free will, let me give another example which demonstrates that free will does not play a major role in shaping the future, nor does the predictability of the future have an impact on free will. The greater metropolitan area of Athens currently contains about 4 million people, more than 40 percent of Greece's population. In the beginning of this century the area consisted mostly of two towns, Athens and Piraeus, plus a number of minor villages. It grew as masses drifted in from all over the country and occupied makeshift housing on the town's perimeter. From time to time the city limits would expand to include the new inhabitants. No city planning was ever attempted, which is obvious from a bird's-eye view. Nevertheless, the city functions, accommodates hordes of tourists, and remains a favorite place for the majority of Athenians, a fact that motivated the Athens Institute of Ekistics (studies of human settlement) to carry out a major study on this apparently chaotic form of community development.

The researchers documented and classified all the services provided by the city (taverns, movies, grocery shops, railroad stations, opera house, and so forth). In so doing they discovered within the city as a whole a structure of many small communities defined by the common use of most basic facilities. They also discovered that the total person-kilometers for each one of these communities was about the same. If population density increased in one community—and consequently the total person-kilometers—new facilities were created and the community split into two. The balance at any time was between "paying" for traveling and paying for creating new facilities. It was all done in such a way as to optimize the expenditure of energy, the biological equivalent of money.

Obviously, for facilities used rarely (such as a theater, swimming pool, and park), one would be willing to walk further; these types of services were present in only one out of seven communities, but also serviced the six closest ones. This hierarchy continued upward, so that for services used even more rarely (such as a stadium, concert hall, and museum) one would have to travel even further, and more communities would share the same facilities. Interestingly, the ratio *seven* was maintained between hierarchical levels. Furthermore, the study identified a total of five such hierarchical levels nested like Russian dolls.

A question addressed by J. Virirakis[10] was how well such a structure optimized energy expenditure at the city level, five hierarchical levels higher than where it was first detected. He wrote a computer program to calculate energy expenditure and ran it for different configurations of the distribution of services. The simplest configuration—which would have appealed only to naive city planners—was that all facilities are concentrated in the middle of the city. Another one was to have facilities distributed at random throughout the city. The first case gave an increase in energy spent of a factor of six; the second a factor of fifteen over the actual configuration.

The city of Athens has been optimized naturally according to a simple energy-saving principle. The overall structure is highly ordered and scientifically describable but does not come in conflict with individual free will. Every Athenian can buy bread at any bakery in town, but most probably he or she will buy it at the nearest one. This is the key that makes the system work.

Free will comes under different light in another case of optimization, the closed-circuit car race. Consider a Formula One closed-circuit race. Without too much difficulty a scientifically minded person can write a computer program to optimize the driver's decisions during the race. Among the data needed are the power of the car, the ratio of the gears, the total weight, the coefficient of friction between wheels and pavement, and the detailed circuit. Then some laws of physics must be built in: centrifugal forces, accelerations, and the like. That done, a printout can be produced that dictates the actions a driver must take to cover one hundred laps of the circuit in the shortest possible time. After the race one may confront the winner with this action list and ask if he or she carried them out. The winner will claim, of course, that he or she was acting with free will but would have to agree that what is on the paper represents what was done; otherwise the winner would not have won.

Optimization reduces free choice. From the moment one chooses to strive for the winning place there is not much freedom left. You must follow the list of optimized course actions as closely as possible. Thus, the winner's actions and decisions are rather predictable, in contrast to a driver who may be accident-prone or a Sunday driver who may decide to stop for an unpredictable reason such as observing a rare bird. Decisions that can be predicted in timing, quality, and quantity are no longer decisions; they are necessities dictated by the role assumed. If the race car driver wants to be a winner, his only real choice is the one of making mistakes. There are many ways to make mistakes but only one way to do it right, and that one can be forecasted.

Many people besides race car drivers consider themselves decision makers. A better name for them would be *optimizers*. Such a name would also be more appropriate for marketers and companies' top executives who feel a heavy responsibility for making decisions. The gravity they see in their actions causes them, at best, anxiety and, at worst, an ulcer or a heart attack. Yet they behave as optimizers most of the time. Their job is to stay on the course. To do that they need to make corrections, like drivers on a highway who are in fact continuously zigzagging in order to go straight. The best of us may make smaller and less frequent corrections, but none of us is free to make a sharp turn.

Highway driving is not particularly anxiety provoking. Most of the time there is little choice. There is both wisdom and comfort in the whimsical saying, "Rejoice, rejoice, we have no choice." A leader's job to a large extent is to optimize; that is, reduce the amplitude and the frequency of the corrections to be applied. The burden of such responsibility is not unbearable. If "decision makers" became more aware of well-established natural-growth processes and of how much free choice they may not have after all, they would benefit not only from reduced stress but also from the avoidance of mistakes.

EPILOGUE

• • •

We are in Central Europe. I am driving with my son on a highway flanked by towering Alps on one side and a lake reflecting the setting sun on the other. Neither of us is talking. He is fifteen, and if I had to describe with one word what he is going through, it would be learning.

We are listening to Chopin's nocturnes reproduced with rich sound on the car's four-way stereo. Music, scenery, and motion blend together to evoke a contemplative mood. Finally, my son interrupts the music. "Why did Bartok become successful?" he asks. "His music is not nearly as effective as Chopin's."

I am accustomed to such questions. I want to respond quickly and meaningfully. I "dance" my way through the answer: "Bartok made innovations in music. All composers do. But his innovations matched most closely the innovation assimilation rate of the music lovers of his time. Had Chopin composed Barok's kind of music, it would have been a fiasco, for no other reason than timing. Evolution in music follows a natural course, and a composer will fail to reach an audience by going faster or slower."

It all came out of me in one shot and sounded right, so I carried it further. "What you could have asked is why there are more Chopin fans than Bartok fans. Well, natural evolution follows S-curves, and that is what classical music did. It was born, let's say, sometime in the fifteenth century and then grew in complexity,

223

importance, innovation, and appreciation. In the twentieth century
it started approaching its ceiling. In between, the eighteenth century, the rate of growth went over a maximum. Around that time
composers' efforts were rewarded more handsomely than today.
Gifted composers today are given limited space. If they innovate,
they find themselves quickly above the public's acceptance curve,
which has flattened out; if they don't, they are not saying anything
new. In neither case will they be credited with an achievement."

• • •

S-curves have transcended the space of professional and academic interest for me; they have become a way of life. The same thing happened
to other "tools" that I received in my training. The oldest one is the
scientific method summarized as observation, prediction, verification.
This sequence of actions permits the use of the stamp "scientific" on a
statement, which enjoys widespread respect, but most important helps
the one who makes the statement become convinced of its validity.

Another "tool" is evolution through natural selection, which can also
be reduced to three words: mutation, selection, diffusion. Mutations owe
their existence to the law which says that when something can happen,
it will happen. Mutations serve as emergency reserves; the larger their
number, the higher the chance for survival. (Conglomerates stay in business longer than specialized companies.) The selection phase is governed
by competition, which plays a supreme role and deserves to be called the
"father of everything." After selection, the diffusion of the chosen mutant proceeds along natural-growth curves, smoothly filling the niche to
capacity. Irregular oscillations toward the end of a growth curve may be
heralds of a new growth phase that will follow.

Fads and media stories about certain issues rise and subside, and in so
doing produce waves of interest. These waves reflect the public's preoccupation, but at the same time they stimulate it, thus mixing cause and
effect. The feedback loop does not deteriorate into a vicious circle that
goes on forever, however. Fads and preoccupations have a certain potential to exhaust and a life cycle to go through. Each wave of interest
replicates a natural-growth process.

Once growth is complete, the level reached reflects an equilibrium. Its
signature becomes an invariant or constant that, despite erratic fluctuations, manifests the existence of tolerance thresholds and social balance.
To find these constants one may have to look beyond the sensational
headlines, which sometimes mask the essential message. Social living is

self-regulated and may ultimately defy legislation and public opinion.

Finally, there is the "tool" of the overall cycle that clocks people's whims and adventures with a period of about fifty-six years. Like a slow underlying pulsation, it rocks society regularly, sending it through waves of violence and destruction, achievement and progress, prosperity and economic depression. All these "tools" can be used quantitatively as means to improve forecasts and decrease the chance of mistakes. Through use of historical data they can help reveal the direction of a course and the extent of its boundaries. One can even estimate the size of the deviations expected above and below the projected path.

But perhaps they can be of even greater help in a qualitative way, without the use of computers, fitting programs, and mathematical calculations. When these tools are grasped more than just intellectually, they give rise to a better understanding of the most probable evolution of a process and how much of it still lies ahead. Such understanding goes well beyond the supertanker analogy, which claims that supertankers cannot make sharp turns, and therefore their immediate course is predictable. The life cycle of natural growth is symmetric, so there is as much to be expected from the time the maximum rate of growth is reached to the end of the cycle as was obtained from the beginning of the cycle to the time of the maximum rate of growth. From half of a growth process one can intuitively predict the other half.

Invariably there is change, some of it imposed, some of it provoked. For one reason or another, transitions always take place. Change may be inevitable, but if it follows a natural course, it can be anticipated and planned for. Timing is important. In the world of business, for example, there is a time to be conservative—the phase of steep growth when things work well and the best strategy is to change nothing. There is also the time of saturation when the growth curve starts flattening out. What is needed then is innovation and encouragement to explore new directions. Our leaders may not be able to do much about changing an established trend, but they can do a lot in preparing, adapting, and being in harmony with it. The same is true for individuals. During periods of growth or transition our attitude should be a function of where we are on the S-curve. The flat parts of the curve in the beginning and toward the end of the process call for action and entrepreneurship, but the steeply rising part in the middle calls for noninterference, letting things happen under their own momentum.

Finally, we can obtain some insight into our longevity. Again, I cite Marchetti as an example. He is more than sixty years old and has already retired one and a half times (he is now working half-time), but never-theless he is producing papers at a frantic rate. The last time I saw him he told me that he had a backlog of about twenty-five articles to write. Death rarely strikes those who are enjoying maximum productivity, and rumor says that Marchetti is overproducing on purpose in order to live longer. I asked him if he had computed his curve.

"Yes, I have," he said and paused. Then he added, as if in silent assent to the correlation between productivity and life span, "I look both ways before crossing the street."

For me, his real motivation for being productive is irrelevant. If his output does not yet show any signs of decline, I feel confident that he will continue to live for a long time to come.

APPENDICES

APPENDIX A

Mathematical Formulation of S-Curves and the Procedure for Fitting Them on Data

The behavior of populations growing under Darwinian competition is described in general by the celebrated Volterra-Lotka equations. These equations—one for every competing species—specify the rate of growth which is made up of several terms.

$$\frac{dN_i}{dt} = a_i N_i - \sum_{j=1}^{n} b_{ij} N_i N_j$$

The positive term is proportional to the actual size. The parameter a_i reflects the capacity of the species to reproduce, or, in other words, the rate of growth of population i in the absence of predators. The negative terms represent growth reduction due to loss of life caused by the predator. The parameter b_{ij} can be interpreted as the rate at which predator j will grow assuming there is an infinite amount of prey i. The properties of the solutions to these equations have been described by Elliott Montroll and N. S. Goel[1] and more recently by M. Peschel and W. Mendel.[2] We can distinguish the following special cases.

The Predator-Prey Equations

Two species are locked together in a life-death relationship because one serves as food for the other; for example, lynx-hare, or big fish–small fish.

The populations feature oscillations. The mathematical description is obtained from the equation:

$$\frac{dN_1}{dt} = a_1 N_1 - k_1 N_1 N_2 \qquad\qquad\qquad \text{prey}$$

$$\frac{dN_2}{dt} = -a_2 N_2 + k_2 N_1 N_2 \qquad\qquad\qquad \text{predator}$$

where N_1 and N_2 are the populations of prey and predator respectively. The k constants represent the strength of the interaction between the two species. More mathematical details can be found in Montroll and Goel's work.[3]

The Malthusian Case: One Species Only

An illustrative example of this case is a population of bacteria growing in a bowl of broth. The bacteria act as the agent that transforms the chemicals present in the broth into bacteria. The rate of this transformation is proportional to the number of bacteria present and the concentration of transformable chemicals.

All transformable chemicals will eventually become bacteria. One can therefore measure broth chemicals in terms of bacterial bodies. If we call $N(t)$ the number of bacteria at time t, and M the amount of transformable chemicals at time 0 (before multiplication starts), the Verhulst equation can be written as

$$\frac{dN}{dt} = aN (M - N) \tag{1}$$

and its solution is

$$N(t) = \frac{M}{1 + e^{-(at+b)}} \tag{2}$$

with b a constant locating the process in time.[4]

We can manipulate mathematically Equation (2) in order to put it in the form

$$\frac{N(t)}{M - N(t)} = e^{(at+b)} \tag{3}$$

Taking the logarithm of both sides, we obtain a relationship linear with time, thus transforming the S-shaped curve of Equation (2) into the straight line of equation: $at + b$. The numerator of the left side of Equation (3) is the new population while the denominator indicates the space still remaining empty in the niche. In the case of one-to-one substitutions (see below), the numerator is the size of the *new* while the denominator is the size of the *old* at any given time. If on the vertical logarithmic scale we indicate the fractional market share instead of the ratio *new/old,* we obtain the so-called logistic scale in which 100 percent goes to plus infinity and 0 percent to minus infinity.

M is often referred to as the niche capacity, the ceiling of the population $N(t)$ at the end of growth. In order to solve Equation (1), M must be constant throughout the growth process, but this requirement can be relaxed for one-to-one substitutions.

One-to-One Substitutions: Two Competitors in a Niche

This case originally treated by J. B. S. Haldane[5] and later reported by Alfred Lotka[6] deals with the situation where a mutant, that is, a new variety that we can call N_2, has a small genetic advantage k over the old variety N_1. This means that at every generation the ratio of individuals in the two varieties will be multiplied by $1/(1-k)$. In n generations, then, we will have the ratio of populations as

$$\frac{N_2}{N_1} = \frac{R_o}{(1-k)^n} \text{ where } R_o = \frac{N_2}{N_1} \text{ at } t=0. \tag{4}$$

Whenever k is small—in biology it is typically around 0.001—Equation (4) can be approximated as

$$\frac{N_2}{N_1} = R_o \, e^{kn} \tag{5}$$

which is the same as Equation (3) since $N_2 = M - N_1$. The only difference between this case and the one in the previous section is that here we have an initial condition (R_o) while there we had a final condition (M). In a typical application the single occupancy implies determination of the final ceiling M from early data, a procedure subject to large uncertainties addressed in detail in Appendix B. The two-competitor typical case deals with evolution in *relative* terms. The final ceiling of the process is by definition equal to 100 percent. The determination of the trajectory is not sensitive to the size of the niche, which may be varying throughout the substitution process. This permits a reliable forecast of the competitive "mix" even in situations where the overall "market" may be going through unpredictable convulsions.

Multiple Competition: Many Competitors at the Same Time

One approach dealing with the many-competitor case developed by Nebajsa Nakicenovic[7] is based on successive one-to-one substitutions. The ratio of population i divided by the sum of all other populations follows Equation (3) for all competitors except one, typically the leader, whose share is calculated as what remains from 100 percent after all other shares have been subtracted. Thus, the trajectory of every competitor will in general be an S-curve—either a growing one or a declining one—except during the saturating phase; that is, while in transition between the end of the phasing in and the beginning of the phasing out.

The generalized equations for the market shares of n competitors are:

$$\left. \begin{array}{c} f_i\,(t) = \dfrac{1}{1 + e^{-(at+b)}} \\[3ex] f_j(t) = 1 - \displaystyle\sum_{i=1}^{n} f_i(t) \end{array} \right\} \quad \text{for } i{=}j$$

and

Fitting an S-Curve on a Set of Data Points

There are many ways to fit a mathematical function onto a sequence of data points. The method used extensively in this book, and the one likely to produce the best results, involves iterations through a computer program that tries to minimize the following sum

$$\sum_i \frac{(F_i - D_i)^2}{W_i}$$

where F_i is the value of the function, D_i is the data value, and W_i is the weight we may want to assign, all at time t_i. The function is originally evaluated unintelligently by assigning arbitrary starting values to the parameters of F, but the program performs a trial-and-error search through many iterations to determine those values for which the sum becomes as small as possible.

For S-curve fitting, the function F takes the form of Equation (2), while for one-to-one substitutions it takes the form (3) or simply the straight line expression $at + b$.

By assigning weights to the different points we can force the computer to "pay attention" to some data points more than others, if there are doubts about the validity of the data.

The program we used proved quite robust in the sense that changing weights and starting values affected only the computing time but not the final results. Still, we found variations and correlations between the parameters a, b, and M, particularly in the instance of the Malthusian case. In Appendix B we present some results of an extensive study on the error sizes to be expected and the correlations between the three parameters of an S-curve determined from a fit.

APPENDIX B

Expected Uncertainties on S-Curve Forecasts

Alain Debecker and I undertook an extensive computer simulation (a Monte Carlo study) to quantify the uncertainties on the parameters determined by S-curve fits.[1] We carried out a large number of fits (around forty thousand), on simulated data randomly deviated around theoretical S-curves and covering a variety of time spans across the width of the life cycle. We fitted the scattered data using Equation (2) with the constant b replaced by at_o so that t_o has the units of time. Each fit yielded values for the three parameters M, a, and t_o. With many fits for every set of conditions, we were able to make distributions of the fitted values and compare their average to the theoretical value used in generating the scattered data. The width of the distributions allowed us to estimate the errors in terms of confidence levels.

The results of our study showed that the more precise the data and the bigger the section of the S-curve they cover, the more accurately the parameters can be recovered. I give below three representative look-up tables. From Table II we can see that historical data covering the first half of the S-curve with 10 percent error per data point will yield a value for the final maximum accurate to within 21 percent, with a 95 percent confidence level.

As an example, let us consider the sales of the minicomputer VAX 11/750 shown in Figure 3.1. At the time of the fit there had been 6,500 units sold and M was estimated as 8,600 units. Consequently, the S-curve section covered by the data was $6{,}500/8{,}600 = 76$ percent. From the scattering of the trimesterly sales, the statistical error per point, after accounting for seasonal variations, was evaluated as 5 percent. From Table III, then, we obtained the uncertainty on M as somewhat higher than 4 percent for a confidence level of 95 percent. The final ceiling of 8,200 fell within the estimated uncertainty.

Finally, we were able to establish correlations between the uncertainties on the parameters determined. One interesting conclusion was that among the S-curves that can all fit a set of data, with comparable statistical validity, the curves with smaller values for a have bigger values for M. In other words, a slower rate of growth correlates to a larger niche size, and vice versa. This implies that accelerated growth is associated with a lower ceiling, bringing to mind such folkloric images as short life spans for candles burning at both ends.

TABLE I

Expected uncertainties on M fitted from data covering the range 1 percent to 30 percent of the total S-curve. The confidence level is marked vertically, while the error on the historical data points is marked horizontally. All numbers are in percentages.

	1	5	10	15	20	25
70	2.7	13	28	47	69	120
75	3.2	15	32	53	81	190
80	3.9	17	36	62	110	240
85	4.8	19	41	81	130	370
90	5.9	22	49	110	210	470
95	48.0	49	180	350	690	

TABLE II

Expected uncertainties on M fitted from data covering the range 1 percent to 50 percent of the total S-curve. The confidence level is marked vertically, while the error on the historical data points is marked horizontally. All numbers are in percentages.

	1	5	10	15	20	25
70	1.2	5.1	11	17	23	29
75	1.4	5.5	12	19	26	32
80	1.8	6.4	14	22	29	36
85	2.1	7.3	16	25	36	42
90	2.6	8.8	18	29	42	48
95	3.1	11.0	21	39	56	66
99	4.6	22.0	30	55	150	110

TABLE III

Expected uncertainties on M fitted from data covering the range 1 percent to 80 percent of the total S-curve. The confidence level is marked vertically, while the error on the historical data points is marked horizontally. All numbers are in percentages.

	1	5	10	15	20	25
70	0.5	1.9	3.9	5.1	8.1	8.9
75	0.6	2.1	4.4	5.5	9.0	9.6
80	0.7	2.4	4.8	6.2	9.8	11.0
85	0.8	2.8	5.5	7.1	12.0	13.0
90	1.1	3.3	6.3	9.1	13.0	16.0
95	1.3	4.0	7.6	11.0	16.0	18.0
99	2.2	5.6	9.1	15.0	21.0	31.0

APPENDIX C

Additional Figures

THE NORMAL DISTRIBUTION IS VERY SIMILAR TO THE NATURAL LIFE CYCLE

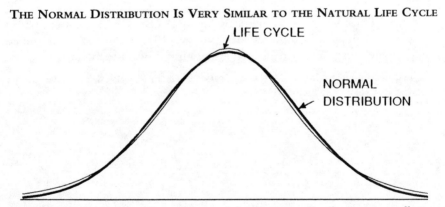

APPENDIX FIGURE 1.1   Comparison of the Gaussian distribution—usually referred to as *normal*—and a natural life cycle.

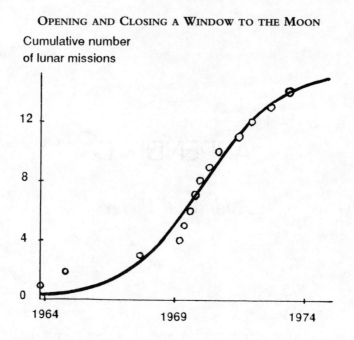

OPENING AND CLOSING A WINDOW TO THE MOON

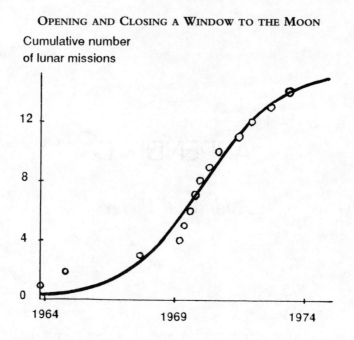

APPENDIX FIGURE 2.1   The cumulative number of moon explorations traced out a rather complete S-curve between 1964 and 1972. Manned and unmanned launches are considered alike as long as the mission was the moon.*

* The data come from the *World Almanac & Book of Facts,* 1988 (New York; Newspaper Enterprise Association, Inc., 1984).

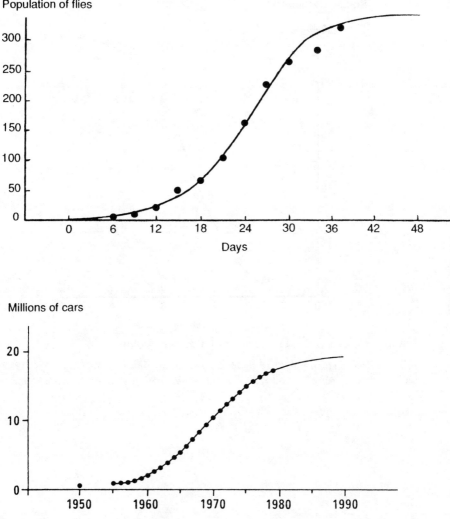

APPENDIX FIGURE 3.1   Data points and fitted curves for 2 populations. At the top, the population of drosophila (fruit flies) under controlled experimental conditions. At the bottom, the number of registered cars in Italy.*

* The top graph was first published by R. Pearl and S. L. Parker, *American Naturalist,* vol. 55 (1921): 503; vol. 56 (1922): 403, as cited by Alfred J. Lotka, *Elements of Physical Biology,* (Baltimore, MD: Williams & Wilkins Co., 1925). The bottom graph was published by C. Marchetti, "The Automobile in a System Context: The Past 80 Years and the Next 20 Years," *Technological Forecasting and Social Change,* vol. 23 (1983): 3–23. Copyright 1983 by Elsevier Science Publishing Co., Inc. Reprinted by permission of the publisher.

THE SIZE OF AN ORGANISM GROWS LIKE A POPULATION

Square centimeters

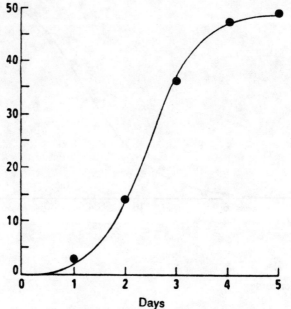

APPENDIX FIGURE 3.2   The growth of a bacteria colony.*

* Adapted from a graph published by H. G. Thornton, *Annals of Applied Biology,* 1922, p. 265, as cited by Alfred J. Lotka, *Elements of Physical Biology* (Baltimore, MD: Williams & Wilkins Co., 1925).

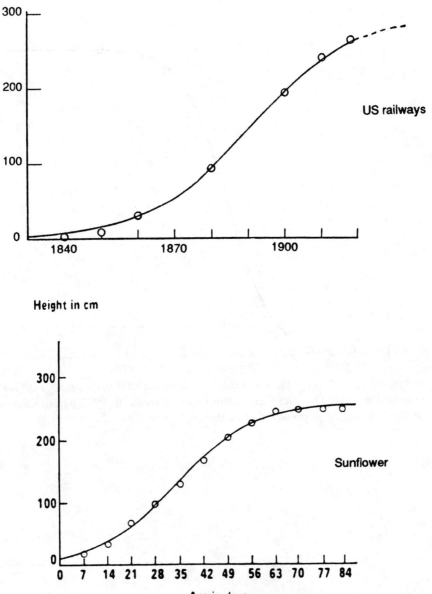

RAILWAY NETWORKS GROW LIKE PLANTS

Thousands of miles

US railways

Height in cm

Sunflower

Age in days

APPENDIX FIGURE 3.3   The graph at the top represents the total mileage of railways in the United States. At the bottom we see the growth in height of a sunflower seedling.*

* The two graphs were published by Alfred J. Lotka, *Elements of Physical Biology* (Baltimore, MD: Williams & Wilkins Co., 1925). The bottom graph is attributed to H. S. Reed and R. H. Holland, *Proc. Natl. Acad. Sci.,* vol. 5 (1919); 135–44.

SUPERTANKERS: THE CRAZE OF A DECADE

Cumulative number of
supertankers constructed

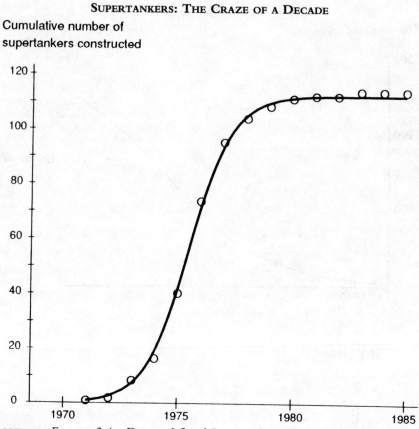

APPENDIX FIGURE 3.4   Data and fitted S-curve for all ships over 300,000 tons
ever constructed. The data were communicated to me in 1985 by Jan Olafsen
of the Norwegian shipbrokers' firm Bassoe A/S & Co.

GOTHIC CATHEDRALS

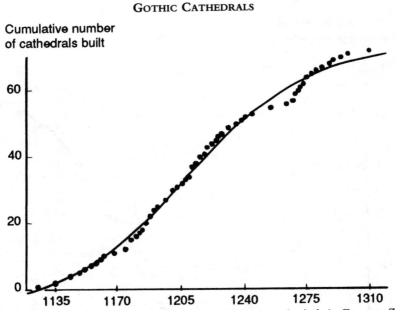

APPENDIX FIGURE 3.5   The construction of Gothic cathedrals in Europe. The dates refer to the beginning of construction. The fitted curve describes the process fairly well.*

* The data come from L. Cloquet, *Les cathédrales et basiliques latines, byzantines et romanes du monde catholique* (Roman and Byzantine Cathedrals and Basilicas of the Catholic World). (Paris: Desclée, De Brouwer et Cie., 1912).

THE ERA OF PARTICLE ACCELERATORS

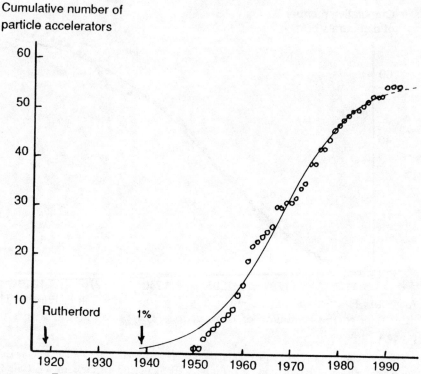

APPENDIX FIGURE 3.6 The data on particle accelerators coming into operation worldwide were fitted with an S-curve whose nominal beginning points at World War II rather than Rutherford's famous experiment.*

*The data come from Mark Q. Barton, *Catalogue of High Energy Accelerators*, Brookhaven National Laboratory, BNL 683, 1961, and J. H. B. Madsen and P. H. Standley, *Catalogue of High-Energy Accelerators* (Geneva: CERN, 1980, plus updates).

JOHANNES BRAHMS (1833–1897)

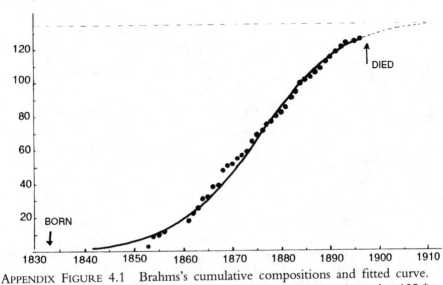

Cumulative number
of compositions

APPENDIX FIGURE 4.1   Brahms's cumulative compositions and fitted curve.
The curve has its nominal beginning in 1843. The ceiling is estimated as 135.*

* The data come from Claude Rostand, *Johannes Brahms* (Paris: Fayard, 1978).

ERNEST HEMINGWAY (1899–1961)

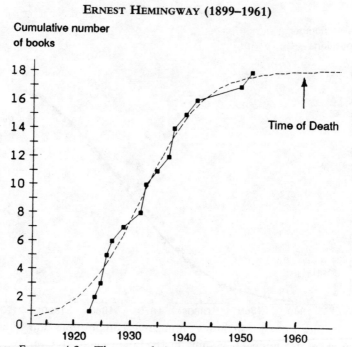

APPENDIX FIGURE 4.2   The cumulative number of books published by Hemingway. The fitted curve starts in 1909 and has a ceiling of 18.*

* The data come from Philip Young and Charles W. Mann, *The Hemingway Manuscripts: An Inventory,* University Park, PA, and London: Pennsylvania State University Press, 1969. Also from Audrey Hanneman, *Ernest Hemingway: A Comprehensive Bibliography* (Princeton, NJ: Princeton University Press, 1967).

PERCY B. SHELLEY (1792–1822)

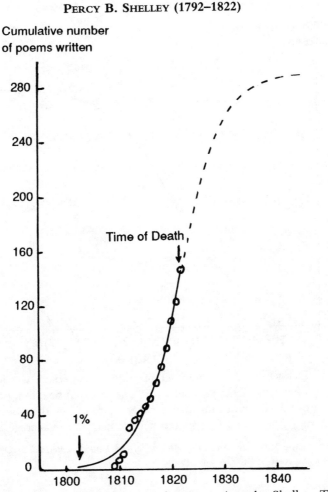

APPENDIX FIGURE 4.3   The number of poems written by Shelley. The dates refer to the time of writing. The fitted curve's nominal beginning is around 1802. The estimated ceiling is 290, about twice the number of works completed at the time of his death.*

* The data come from Newman Ivey White, *Shelley,* vols. 1 and 2 (London: Secker & Warburg, 1947).

THE BIRTH RATE OF AMERICAN WOMEN

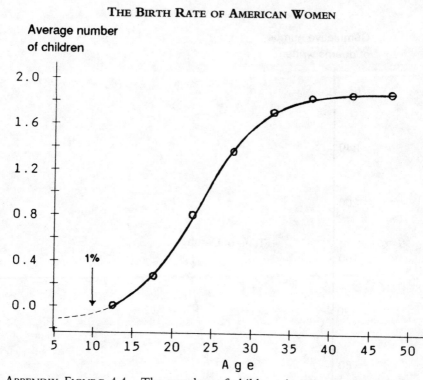

APPENDIX FIGURE 4.4   The number of children American women have on average as a function of the woman's age. The nominal beginning of the fitted curve is around 10, indicating that fertility starts earlier than observed. The ceiling during 1987 was 1.87 children per woman on the average.*

* The data come from the *Statistical Abstract of the United States,* U.S. Department of Commerce, Bureau of the Census, 1989.

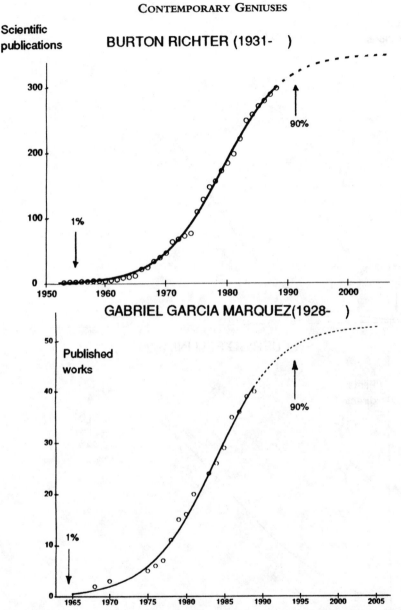

APPENDIX FIGURE 4.5   The creativity curves of four contemporary individuals. For each case we see the cumulative number of works and the corresponding S-curve as determined by a fit. The dotted lines are extrapolations of the curves. Under ordinary circumstances these people should achieve the 90% levels indicated.

CONTEMPORARY GENIUSES (cont.)

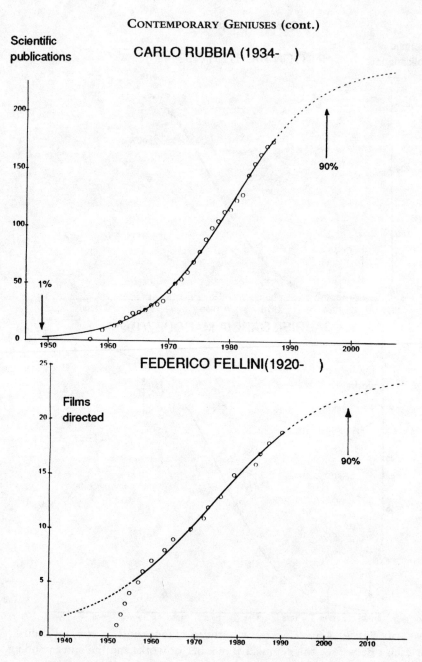

Scientific publications

CARLO RUBBIA (1934-   )

1%

90%

FEDERICO FELLINI(1920-   )

Films directed

90%

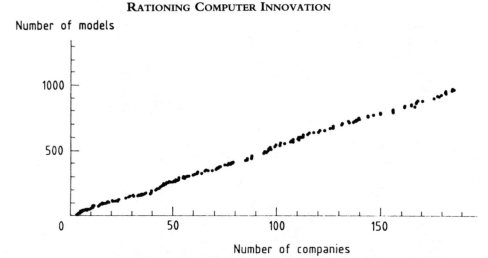

RATIONING COMPUTER INNOVATION

APPENDIX FIGURE 4.6   Over 1,000 data points indicating the total number of different computers versus the total number of different manufacturers that have entered the market. We consider all models above personal computers and all manufacturers, between 1958 and 1985. The points falling on a straight line consistently indicate that there is a strict law in effect demanding 5 models per manufacturer at any given time.*

* The data come from the International Data Corporation's *1985 Processor Data Book,* covering the period from the beginning of 1958 up to the end of 1984. The graph itself has been published in Theodore Modis and Alain Debecker, "Innovation in the Computer Industry," *Technological Forecasting and Social Change,* vol. 33 (1988); 267–78.

LIVES CLAIMED BY THE RED BRIGADES

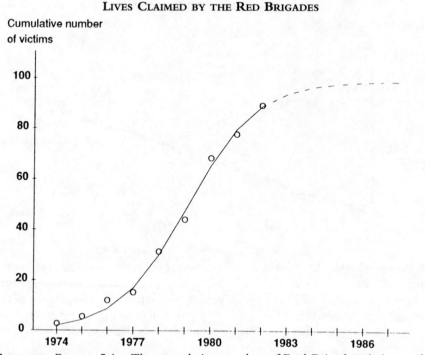

APPENDIX FIGURE 5.1   The cumulative number of Red Brigades victims and a fit. The estimate of the ceiling is 99. The group's activities claimed 90% of it.*

*This figure has been adapted from one by Cesare Marchetti with a nonlinear scale; see Cesare Marchetti, "Intelligence at Work: Life Cycles for Painters, Writers, and Criminals," invited paper to the Conference on the Evolutionary Biology of Intelligence, organized by the North Atlantic Treaty Organization's Advanced Study Institute, Poppi, Italy, July 8–20, 1986, a report of the International Institute of Advanced Systems Analysis, Laxenburg, Austria.

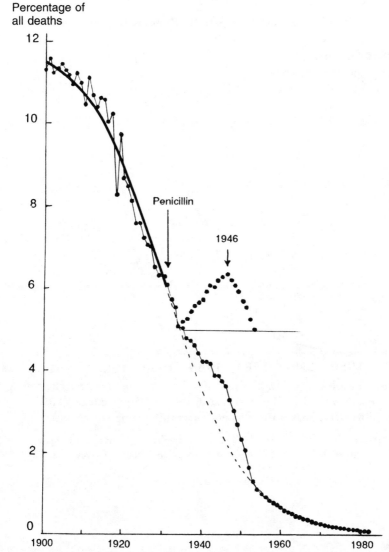

Percentage of
all deaths

Penicillin

1946

APPENDIX FIGURE 5.2   Tuberculosis victims in the United States. The fit is only on the historical window 1900–31. Penicillin was discovered in 1931, but effective medication that may be considered a "miracle drug" for this case became available in 1955. Ironically, the agreement between the data and the curve determined from the first three decades becomes best for the years *after* 1955! A significant excess seems to be related to World War II. In the inset I have extracted the contents of the bump by subtracting from the actual data the share estimated by the smooth line.*

* The data come from the *Historical Statistics of the United States, Colonial Times to 1970,* vols. 1 and 2, Bureau of the Census, Washington, DC, 1976, and from the *Statistical Abstract of the United States,* U.S. Department of Commerce, Bureau of the Census, 1986–91.

**THE AIDS NICHE MAY ALREADY BE PRACTICALLY FILLED IN THE
UNITED STATES**

**AIDS victims as a
percentage of all deaths**

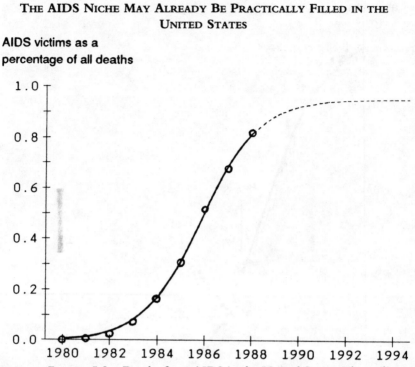

APPENDIX FIGURE 5.3   Deaths from AIDS in the United States. The ceiling of
the fitted S-curve is 0.95. The fact that in 1988, 0.85 percent of all deaths were
attributed to AIDS makes this "niche" practically complete.*

* The data have been provided by HIV/AIDS Surveillance, Centers for Disease Control,
U.S. Department of Health and Human Services, Atlanta, GA; issued January 1990.

Steam engines as a
percentage of all locomotives

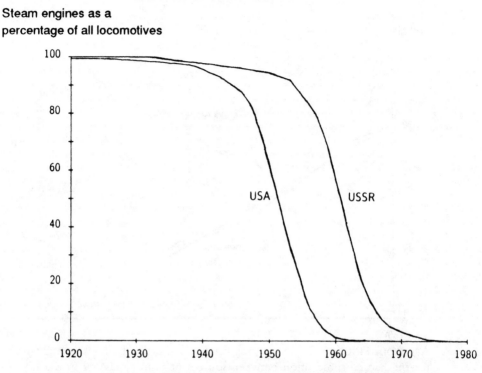

APPENDIX FIGURE 6.1   S-curve fits would seem superfluous to this data showing the decline of steam locomotives in the United States and the U.S.S.R. The
complementary rising S-curves for the percentages of the new "species," diesel
plus electric locomotives, are not shown.*

* Adapted from a graph by Arnulf Grubler in *The Rise and Fall of Infrastructures,* 1990.
Reprinted by permission of the publisher, Physica-Verlag, Heidelberg.

FOSSIL-BASED ENERGY SOURCES REPLACED RENEWABLE ENERGY SOURCES

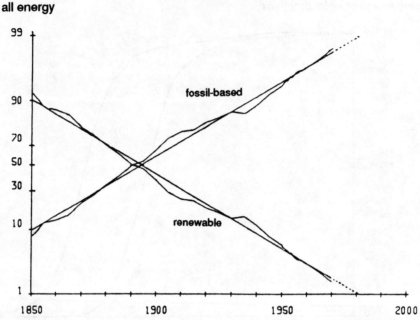

APPENDIX FIGURE 6.2   Data (irregular lines) and fitted curves (straight lines) for the natural substitution between sources of primary energy in the United States. The straight lines are S-curves transformed by the logistic vertical scale. Fossil-based energies are coal, oil, gas, and nuclear. Renewable sources are wood, wind, water, and animal feed.*

* Adapted from a graph by Nebojsa Nakicenovic in "The Automobile Road to Technological Change: Diffusion of the Automobile as a Process of Technological Substitution," *Technological Forecasting and Social Change,* vol. 29 (1986): 309–40. Copyright 1986 by Elsevier Science Publishing Co., Inc. Reprinted by permission of the publisher.

WE ARE BECOMING INFORMATION WORKERS

Percentage of
all workers

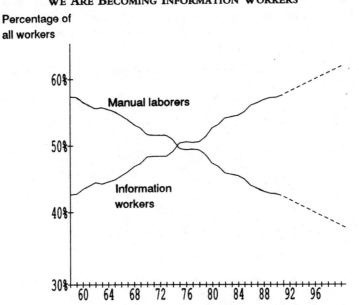

WOMEN IN EXECUTIVE ROLES REACH EQUALITY WITH MEN
IN THE YEAR 2000

Percentage of
all executives

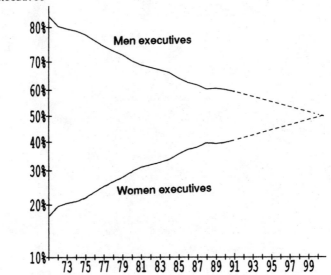

APPENDIX FIGURE 6.3 (page 257)   Two substitutions displaying a natural (straight line) character. Above, the growing information content in jobs displaces manual labor. Below, the percentage of women among executives grows to reach that of men by the year 2000.*

* The data come from *Employment and Earnings,* Bureau of Labor Statistics, U.S. Department of Labor, 1980–91.

THE DESTRUCTION OF THRESHING MACHINES

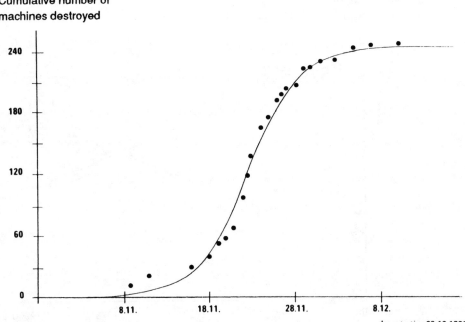

APPENDIX FIGURE 6.4   Cumulative attacks on threshing machines during 1830 in England by farm laborers opposing the new technology. The wave of destruction lasted only one month but proceeded "naturally" as evidenced by the S-shaped curve.*

* Adapted from a graph by Cesare Marchetti in "On Society and Nuclear Energy," Commission of the European Communities, report EUR 12675 EN, Luxemburg, 1990. Reprinted by permission of the Commission of the European Communities.

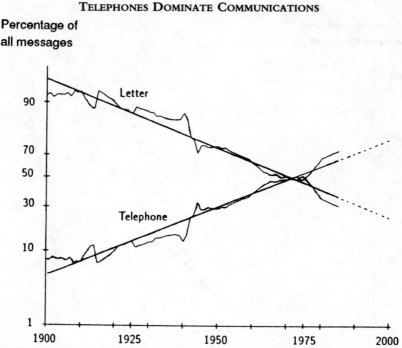

TELEPHONES DOMINATE COMMUNICATIONS

Percentage of
all messages

APPENDIX FIGURE 6.5   Substitution of letters by telephone calls in the total number of messages communicated in France between 1900 and 1985. Equal numbers of messages were exchanged by each medium around 1970. The irregular lines are the actual data while the idealized substitution trajectories have become straight lines thanks to the logistic scale.*

* Adapted from a graph by Arnulf Grubler in *The Rise and Fall of Infrastructures*, 1990. Reprinted by permission of the publisher, Physica-Verlag, Heidelberg.

WARS MAY INTERFERE WITH NATURAL SUBSTITUTIONS

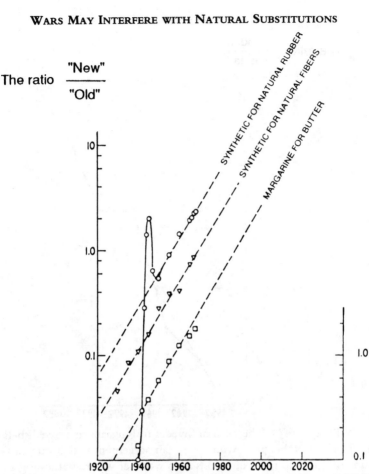

APPENDIX FIGURE 6.6   Three replacements in the United States published by
J. C. Fisher and R. H. Pry in 1971. The data show ratios of amounts consumed.
The logarithmic scale brings out the straight-line character of the substitutions.
The only significant derivation from a *natural* path is observed for the production
of synthetic rubber during World War II.*

* Adapted from a graph by J. C. Fisher and R. H. Pry in "A Simple Substitution Model
of Technological Change," *Technological Forecasting and Social Change,* vol. 3 (1971):
75–88. Copyright 1988 by Elsevier Science Publishing Co., Inc. Reprinted by permission of the publisher.

WHY DID SWEDES REPLACE RUSSIANS?

The ratio: $\dfrac{\text{Sweden}}{\text{Russia}}$

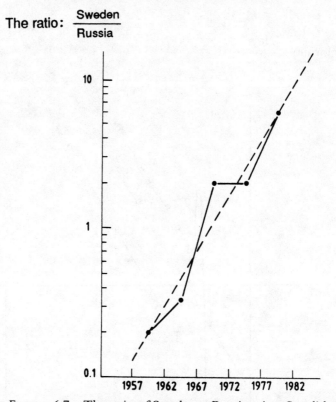

APPENDIX FIGURE 6.7   The ratio of Swedes to Russians in a Swedish-Russian "microniche" of Nobel prize winners. In fair agreement with a straight line the evolution of this ratio raises questions as to whether we are witnessing a natural substitution in passing from an 80% Russian group to an 80% Swedish one 25 years later or whether there was some other factor at work.*

* The figure is reprinted from my article, "Competition and Forecasts for Nobel Prize Awards," *Technological Forecasting and Social Change*, vol. 34 (1988): 95–102.

COMPETITION BETWEEN TRANSPORTATION INFRASTRUCTURES IN THE UNITED STATES

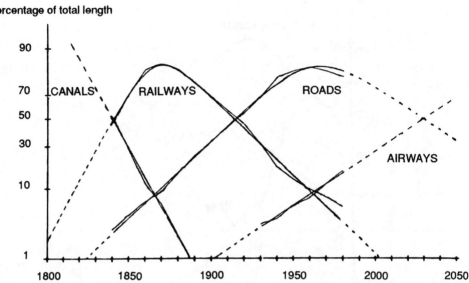

APPENDIX FIGURE 7.1   The sum total in mileage among all transport infrastructures is split here among the major types. A declining percentage does not mean that the length of the infrastructure is shrinking but rather that the total length is increasing. Between 1860 and 1900 the amount of railway track increased, but its share of the total decreased because of the dramatic rise in road mileage. The fitted lines are projected forward and backward in time. The share of airways is expected to keep growing well into the second half of the twenty-first century.*

* Adapted from a graph by Nebojsa Nakicenovic in "Dynamics and Replacement of U.S. Transport Infrastructures," in J. H. Ausubel and R. Herman, eds, *Cities and Their Vital Systems, Infrastructure Past, Present, and Future* (Washington, DC: National Academy Press, 1988). Reprinted by permission of the publisher.

COMPETITION BETWEEN DISEASES

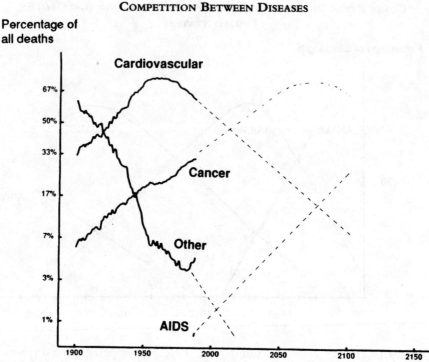

APPENDIX FIGURE 7.2   This figure splits the total number of deaths in the United States into major disease categories. Once again, the vertical scale is logistic. We see only the projections of the straight lines fitted on the data. AIDS in this scenario is given a slope comparable to the one for cardiovascular ailments. The small upward trend of Other in recent years seems important because of the nonlinear scale; it amounts to only a 1% gain on cancer and is most probably due to a random fluctuation.*

* The data come from the *Historical Statistics of the United States, Colonial Times to 1970,* vols. 1 and 2, Bureau of the Census, Washington, DC, 1976, and from the *Statistical Abstract of the United States,* U.S. Department of Commerce, Bureau of the Census, 1986–91.

COMPETITION FOR NOBEL PRIZES

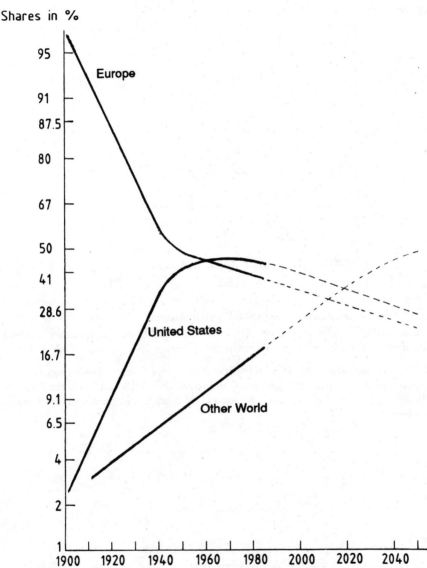

APPENDIX FIGURE 7.3 We see here the split of Nobel prize awards among three major regions in the usual representation. To keep the picture uncluttered, we show only the fitted lines. Complementary straight lines are evidence for three general substitutions: Europeans yielding first to Americans, then to Other, and finally Americans also declining in favor of Other.*

* This figure has been adapted from my article, "Competition and Forecasts for Nobel Prize Awards," in *Technological Forecasting and Social Change*, vol. 34 (1988): 95–102.

ANNUAL ENERGY CONSUMPTION GROWS TO FILL A NICHE

Quadrillions of BTUs

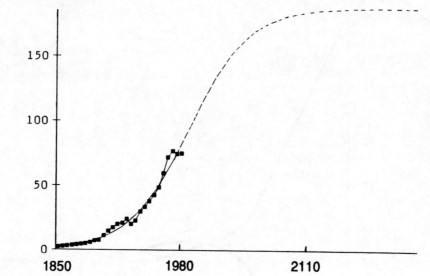

APPENDIX FIGURE 8.1   Data and S-curve fit on annual energy consumption in the United States. Energy is expressed here in quadrillions (one thousand million million) British Thermal Units (BTUs). One BTU is the quantity of heat required to raise by 1° Fahrenheit the temperature of 1 pound of water.*

* The data come from the *Historical Statistics of the United States, Colonial Times to 1970,* vols. 1 and 2, Bureau of the Census, Washington, DC, 1976, and from the *Statistical Abstract of the United States,* U.S. Department of Commerce, Bureau of the Census, 1986–91.

A Slow Cyclical Variation in Sunspot Activity

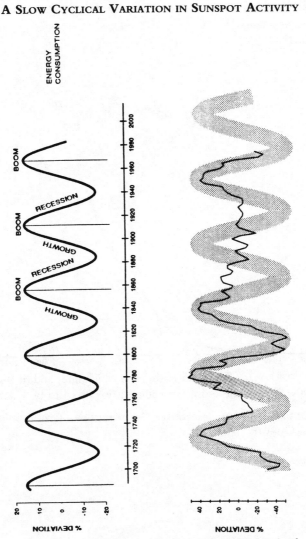

APPENDIX FIGURE 8.2   The lower graph shows the variation in the number of sunspots smoothed over a rolling 20-year period with respect to a 56-year moving average. Such a procedure—routinely used in time series analyses—washes out the well-known 11-year cycle of sunspot activity and reveals a longer periodic variation similar to the energy-consumption cycle. With one exception—a peak missing around 1900—the oscillation conforms to a 56-year cycle.*

* The data have been obtained from P. Kuiper, ed., *The Sun* (Chicago: The University of Chicago Press, 1953) and from P. Bakouline, E. Kononovitch, and V. Moroz, *Astronomie Générale*. Translated into French by V. Polonski. (Moscow: Editions MIR, 1981).

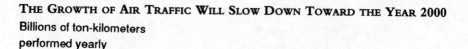

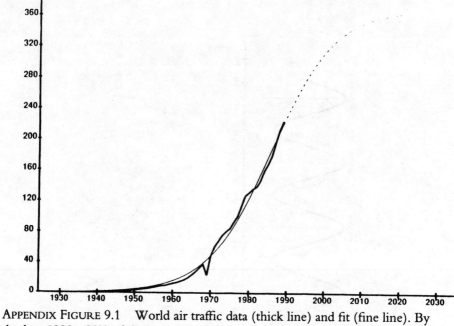

APPENDIX FIGURE 9.1   World air traffic data (thick line) and fit (fine line). By the late 1990s, 90% of the estimated ceiling is expected to be reached.*

* The data for this figure were provided by the International Civil Aviation Organization, Quebec, Canada.

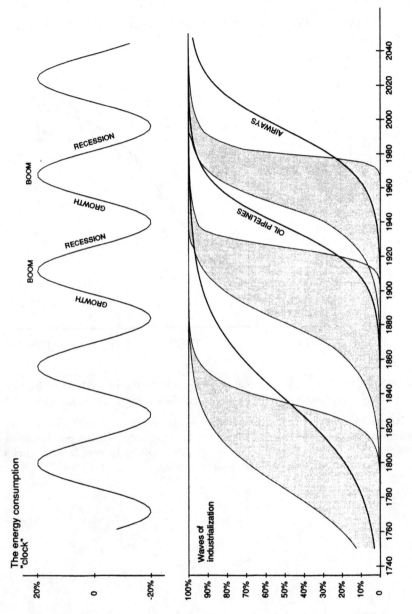

APPENDIX FIGURE 9.2   In a simplified presentation of Figure 9.2 we have added here examples of the rare "tunneling through" process that continues growing unaffected beyond the period of recession. On the first wave it was canal construction in Russia that continued growing. On the present wave, besides the airways shown, computer innovation, natural gas pipelines, nuclear power stations, and pollution abatement also tunnel through. The figure is adapted from Figure 9.2 in the text.*

* The data come mostly from Arnulf Grubler, *The Rise and Fall of Transport Infrastructures*, (Heidelberg: Physica-Verlag, 1990).

Plywood Filled a Niche in the Construction Industry

Billions of
of square feet

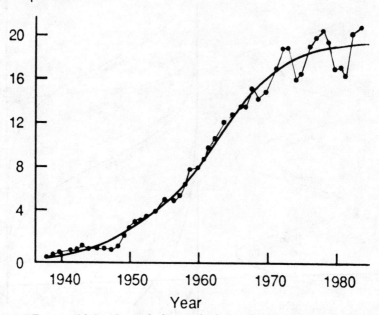

APPENDIX FIGURE 10.1   Annual plywood sales in the United States. Significant deviations from the S-curve appear when the ceiling is approached.*

* Adapted from a figure by Henry Montrey and James Utterback in "Current Status and Future of Structural Panels in the Wood Products Industry," *Technological Forecasting and Social Change*, vol. 38 (1990): 15–35. Copyright 1990 by Elsevier Science Publishing Co., Inc. Reprinted by permission of the publisher.

Natural Growth Alternating with States of Chaos

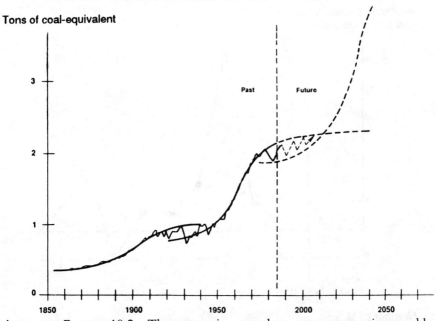

APPENDIX FIGURE 10.2   The per–capita annual energy consumption world-wide: data, fits, and a scenario for the future.*

* This figure reproduces in part a drawing from J. Ausubel, A. Grubler, N. Nakicenovic, "Carbon Dioxide Emissions in a Methane Economy," *Climatic Change,* vol. 12 (1988): 245–63. Reprinted by permission of Kluwer Academic Publishers.

POPULARIZED DEATHS HAVE AN ECHO

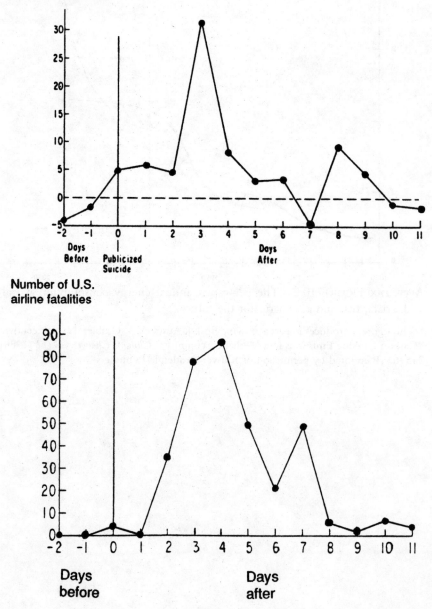

APPENDIX FIGURE 11.1 (opposite)   Both graphs show the number of fatalities in accidents *above* the expected average following a suicide or a homicide of a popular personality. The top graph concerns cars; the lower one, U.S. commercial carriers. There is evidence for a strong 3–4 day echo and a smaller 7–8 day one.*

* Adapted from graphs by David Phillips. The top graph is reprinted by permission from David Phillips, "Suicide, Motor Vehicle Fatalities, and the Mass Media: Evidence Toward a Theory of Suggestion," *American Journal of Sociology,* vol. 84, no. 5: 1150. Copyright 1979 by the University of Chicago. The bottom graph is reprinted by permission from David Phillips, "Airplane Accidents, Murder, and the Mass Media: Towards a Theory of Imitation and Suggestion" *Social Forces,* vol. 58, no. 4 (June 1980). Copyright 1980 by University of North Carolina Press.

WE CAN BECOME MUCH MORE EFFICICENT

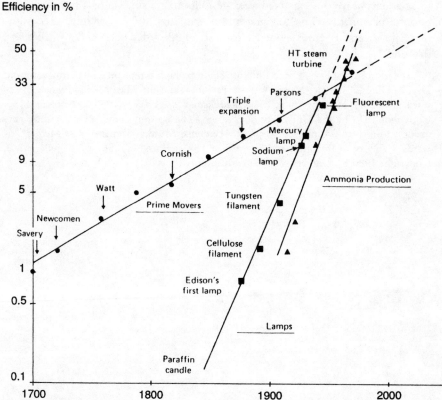

APPENDIX FIGURE 11.2   Historical trends in efficiency for three technologies: engines, production of light, and production of ammonia. The efficiency is defined in the classical thermodynamic way as useful output divided by energy input. In spite of the progress made during the last 300 years, we are still below the 50% efficiency level in these three technologies.*

* Adapted from a graph by Cesare Marchetti in "Energy Systems—the Broader Context," *Technological Forecasting and Social Change,* vol. 14 (1979): 191–203. Copyright 1979 by Elsevier Science Publishing Co., Inc. Reprinted by permission of the publisher.

NOTES AND SOURCES

Prologue

1. Donella H. Meadows, Dennis L. Meadows, Jorgen Randers, William W. Behrens III, *The Limits to Growth* (New York: Universe Books, 1972).
2. Cesare Marchetti, "On 10^{12}: A Check on Earth Carrying Capacity for Man." Report RR-78-7, May 1978, International Institute of Advanced Systems Analysis, Laxenburg, Austria; and in *Energy,* vol. 4, (1979): 1107–17.

Chapter One: Science and Foretelling

1. James Gleick, *Chaos* (New York: Viking, 1988).
2. Spyros Makridakis, et al. "The Accuracy of Extrapolation (Time Series) Methods: Results of a Forecasting Competition," *Journal of Forecasting,* vol. 1, no. 2, (1982): 111–53.
3. As cited in Derek J. de Solla Price, *Little Science, Big Science . . . and Beyond* (New York: Columbia University Press, 1986).
4. Elliott W. Montroll and Wade W. Badger, *Introduction to Quantitative Aspects of Social Phenomena,* (New York: Gordon and Breach Science Publishers, 1974).
5. A similar graph was reported by Cesare Marchetti in 1982. The data in this graph come from the *Statistical Abstract of the United States*, U.S. Department of Commerce, Bureau of the Census; also from the *Historical Statistics of the United States, Colonial Times to 1970,* vols. 1 and 2 (Washington, DC: Bureau of the Census, 1976).
6. John D. Williams, "The Nonsense about Safe Driving," *Fortune,* vol. LVIII, no. 3 (September 1958): 118–19.
7. Patent pending.
8. Β. Βαλαώρας, Α. Χιώλος. Η μεταβαλλομενη δημοπαθολογία των Ελλήνων. (The Changing Causation of Mortality in Greece. Causes of Death). Αιτίες θανάτου, 1960–1985. Ιατρικά Χρονικά, ΙΒ/1, 73–93, 1989.

9. Isaac Asimov, *Exploring the Earth and the Cosmos* (New York: Crown Publishers, 1982).

10. *Good News Bible, Today's English Version,* The Bible Society, Collins/Fontana.

11. J. C. Fisher and R. H. Pry, "A Simple Substitution Model of Technological Change," *Technological Forecasting and Social Change,* vol. 3, no. 1 (1971): 75–88; and M. J. Cetron and C. Ralph, eds., *Industrial Applications of Technological Forecasting* (New York: John Wiley & Sons, 1971).

12. Alain Debecker and Theodore Modis, "Uncertainties in S-curve Logistic Fits," presented at the International Conference on Diffusion of Technologies and Social Behaviour, Laxenburg, Austria, June 14–16, 1989; also at the Sixth International Symposium of Forecasters, Paris, France, May 1986.

13. P. D. Ouspensky, *In Search of the Miraculous* (New York: Harcourt, Brace & World, 1949).

Chapter Two: Needles in a Haystack

1. An S-curve has been fitted on the data found in T. G. Whiston, "Life Is Logarithmic," in J. Rose, ed., *Advances in Cybernetics and Systems* (London: Gordon and Breach, 1974).

2. Theodore Modis, "Learning from Experience in Positioning New Computer Products," *Technological Forecasting and Social Change,* vol. 41, no. 4 (1992).

3. The data here come from *The World Almanac & Book of Facts, 1988* (New York: Newspaper Enterprise Association, Inc., 1987).

4. *Historical Atlas* (Paris: Librairie Académique Perrin, 1987).

Chapter Three: Inanimate Production Like Animate Reproduction

1. P. F. Verhulst, "Recherches mathématiques sur la loi d'accroissement de la population" ("Mathematical Research on the Law of Population Growth"), *Nouveaux Mémoires de l'Académie Royale des Sciences et des Belles-Lettres de Bruxelles,* vol. 18 (1845): 1–40; also in P. F. Verhulst, "Notice sur la loi que la population suit dans son accroissement" ("Announcement on the Law Followed by a Population During Its Growth"), *Correspondance mathématique et physique,* vol. 10: 113–21.

2. E. W. Montroll and N. S. Goel "On the Volterra and Other Nonlinear Models of Interacting Populations," *Review of Modern Physics,* vol. 43(2) (1971): 231. Also, M. Peschel and W. Mendel, *Leben wir in einer Volterra Welt?* (Berlin: Akademie Verlag, 1983).

3. Alfred J. Lotka, *Elements of Physical Biology,* (Baltimore, MD: Williams & Wilkins Co., 1925).

4. See note 3 above.
5. T. W. L. Sanford, "Trends in Experimental High-Energy Physics," *Technological Forecasting and Social Change,* vol. 23: 25–40.
6. Pierre Vilar, *A History of Gold and Money 1450–1920.* (London: NLB, 1976). The data have been updated from the *Statistical Abstract of the United States,* U.S. Department of Commerce, Bureau of the Census; also from the *Historical Statistics of the United States, Colonial Times to 1970,* vols. 1 and 2. (Washington, DC: Bureau of the Census, 1976).
7. *Statistical Abstract of the United States,* and *Historical Statistics of the United States, Colonial Times to 1970,* (see note 6 above).
8. For a compilation of estimates see Cesare Marchetti, "The Future of Natural Gas: A Darwinian Analysis," a presentation given at the Task Force Meeting *"The Methane Age,"* organized jointly by the International Institute of Advanced Systems Analysis and the Hungarian Committee for Applied Systems Analysis, Sopron, Hungary, May 13–16, 1986.
9. Severin-Georges Couneson, *Les Saints nos frères* (The Saints Our Brothers) (Paris: Beauchesne, 1971).
10. These ideas have appeared in an informal note by Cesare Marchetti titled "Penetration Process: The Deep Past," International Institute of Advanced Systems Analysis, Laxenburg, Austria.

Chapter Four: The Rise and Fall of Creativity

1. Cesare Marchetti, *"Action Curves and Clockwork Geniuses,"* International Institute of Advanced Systems Analysis, Laxenburg, Austria, April 1985.
2. Jean and Brigitte Massin, *Wolfgang Amadeus Mozart* (Paris: Fayard, 1978).
3. Mozart's quotation is my translation of a letter excerpt in French cited by the Massins. See note 2 above.
4. Paul Arthur Schilp, *Albert Einstein als Philosoph und Naturforscher* (Braunschweig, Germany: Friedr. Vieweg & Sohn Verlagsgesellschaft, 1979).
5. See note 1 above.
6. The data come from Stanley Sadie, ed. *New Grove Dictionary of Music and Musicians* (London: Macmillan, 1980).
7. Theodore Modis and Alain Debecker, "Innovation in the Computer Industry," *Technological Forecasting and Social Change,* vol. 33 (1988): 267–78.

Chapter Five: Good Guys and Bad Guys Compete the Same Way

1. Theodore Modis, "Competition and Forecasts for Nobel Prize Awards," *Technological Forecasting and Social Change,* vol. 34 (1988): 95–102.
2. Cesare Marchetti, "On Time and Crime: A Quantitative Analysis of the Time Pattern of Social and Criminal Activities," invited paper to the Annual Interpol Meeting, Messina, Italy, October 1985. Also report WP-85-

84, November 1985, International Institute of Advanced Systems Analysis, Laxenburg, Austria.

3. The data come from the *Statistical Abstract of the United States,* U.S. Department of Commerce, Bureau of the Census, 1986–91; also from the *Historical Statistics of the United States, Colonial Times to 1970,* vols. 1 and 2, Bureau of the Census, Washington, DC, 1976.

4. See note 3 above.

5. Henry James Parish, *Victory with Vaccines* (Edinburgh and London: E. & S. Livingstone Ltd., 1968); also Henry James Parish, *A History of Immunization* (Edinburgh and London: E. & S. Livingstone Ltd., 1965).

6. Such a discussion on diphtheria and tuberculosis was first presented by Cesare Marchetti in "Killer Stories: A System Exploration in Mortal Diseases," report PP-82-7, 1982, International Institute of Advanced Systems Analysis.

Chapter Six: A Hard Fact of Life

1. J. C. Fisher and R. H. Pry, "A Simple Substitution Model of Technological Change," *Technological Forecasting and Social Change,* vol. 3, no. 1 (1971): 75–88.

2. Steven Schnaars, *Megamistakes: Forecasting and the Myth of Rapid Technological Change* (New York: The Free Press, 1989).

3. The building is called the Goetheanum and serves as the headquarters of Steiner's Anthroposophical Society. The Goetheanum is in Dornach, close to Basel, Switzerland.

4. Cited by Cesare Marchetti in "On Society and Nuclear Energy," report EUR 12675 EN, 1990, Commission of the European Communities, Luxemburg, Austria.

5. E. J. Hobsbawm and G. Rude, *Captain Swing* (New York: Pantheon Books, 1968).

Chapter Seven: Competition Is the Creator and the Regulator

1. K. Axelos, *Héraclite et la philosophie* (Paris: Les Éditions de Minuit, 1962).

2. Richard N. Foster, *Innovation: The Attacker's Advantage,* (London: Macmillan, 1986).

3. Marc van der Erve, *The Power of Tomorrow's Management,* (Oxford, England: Heinemann Professional Publishing, 1989).

4. Besides the one by Gerhard Mensch, other classifications cited by him that also show the clustering effect include all the significant accomplishments in technical history mentioned by Joseph A. Schumpeter in *Konjunkturzyklen I and II* (Göttingen, Germany: 1961); mixed data (both basic innovations and improvement innovations have been included) from J. Schmookler, *Invention and Economic Growth* (Cambridge: 1966), pp. 220–22.

5. The seasonality of innovations was pointed out by Cesare Marchetti in "Society as a Learning System: Discovery, Invention and Innovation and Innovation Cycles Revisited," *Technological Forecasting and Social Change, vol. 18* (1980): 267–82.

6. Element discovery was first associated with logistic growth in Derek J. de Sola Price, *Little Science, Big Science* (New York, 1967). Elliott Montroll reexamined the situation in 1974 in Elliott W. Montroll and Wade W. Badger, *Introduction to Quantitative Aspects of Social Phenomena* (New York: Gordon and Breach Science Publishers, 1974). Finally, in 1980, Cesare Marchetti hinted that chemical-element discovery may be linked to Kondratieff's cycle; see note 5 above.

7. Donald Spoto, *The Dark Side of Genius: The Life of Alfred Hitchcock* (New York: Ballantine Books, 1983).

8. Nebojsa Nakicenovic, "Software Package for the Logistic Substitution Model," report RR-79-12 (1979), International Institute of Advanced Systems Analysis, Laxenburg, Austria.

9. *Historical Statistics of the United States, Colonial Times to 1970,* vols. 1 and 2, Bureau of the Census, Washington, DC.

10. Cesare Marchetti, "On Transport in Europe: The Last 50 Years and the Next 20," invited paper to the First Forum on Future European Transport, Munich, Germany, September 14–16, 1987.

11. Arnulf Grubler, *The Rise and Fall of Infrastructures* (Heidelberg: Physica-Verlag, 1990), p. 189.

12. Cesare Marchetti, "Primary Energy Substitution Models: On the Interaction between Energy and Society," *Technological Forecasting and Social Change,* vol. 10 (1977): 345–56. This paper was first delivered in Moscow in November 1974 and published in the August 1975 issue of *Chemical Economy and Engineering Review* (CEER).

13. See note 12 above.

14. See note 12 above.

15. As cited by Larry Abraham in his *Insider Report,* vol. VII, no. 9 (February 1990).

16. See note 15 above.

Chapter Eight: A Cosmic Heartbeat

1. Hugh B. Stewart, *Recollecting the Future: A View of Business, Technology, and Innovation in the Next 30 Years,* (Homewood, IL: Dow Jones-Irwin, 1989).

2. Cesare Marchetti first correlated Stewart's energy cycle with Mensch's basic innovations one. For a review of long economic waves, see R. Ayres, "Technological Transformations and Long Waves," parts I and II, *Technological Forecasting and Social Change,* vol. 37, nos. 1 and 2 (1990).

3. Unless otherwise stated, all data used for the graphs of this chapter come from the *Statistical Abstract of the United States,* U.S. Department of Com-

merce, Bureau of the Census; also from the *Historical Statistics of the United States, Colonial Times to 1970,* vols. 1 and 2, Bureau of the Census, Washington, DC.

4. Cesare Marchetti, "Fifty-Year Pulsation in Human Affairs, Analysis of Some Physical Indicators," *Futures,* vol. 17, no. 3 (1986): 376–88. I have obtained similar results for the diffusion of subways with data from Dominique and Mechèle Frémy, QUID 1991 (Paris: Robert Lafont, 1991), p. 1634.

5. A figure like this was first put together by Cesare Marchetti; see note 4 above. The *hay* (U.S.) data come from Nebojsa Nakicenovic, "The Automobile Road to Technological Change: Diffusion of the Automobile as a Process of Technological Substitution," *Technological Forecasting and Social Change,* vol. 29: 309–40.

6. Cesare Marchetti, "The Automobile in a System Context: The Past 80 Years and the Next 20 Years," *Technological Forecasting and Social Change,* vol. 23 (1983): 3–23.

7. See Cesare Marchetti's paper, "Swings, Cycles and the Global Economy," *New Scientist,* no. 1454, May 2, 1985.

8. This explanation is given by Cesare Marchetti; see note 7 above.

9. Alfred Kleinknecht, *Are There Schumpeterian Waves of Innovations?,* research report WP-87-0760, International Institute of Applied Systems Analysis, Laxenburg, Austria, September 1987.

10. J. A. Schumpeter, *Business Cycles* (New York: McGraw-Hill, 1939).

11. R. Ayres, "Technological Transformations and Long Waves," parts I and II, *Technological Forecasting and Social Change,* vol. 37, nos. 1 and 2 (1990).

12. N. D. Kondratieff, "The Long Wave in Economic Life," *The Review of Economic Statistics,* vol. 17 (1935):105–115.

13. Cesare Marchetti, "On Energy Systems in Historical Perspective: The Last Hundred Years and the Next Fifty," internal note, International Institute of Applied Systems Analysis.

14. See Cesare Marchetti in note 13 above.

15. Simon van der Meer, private communication.

16. Gerald S. Hawkins, *Stonehenge Decoded* (London: Souvenir Press, 1966).

17. Tables for the calculation of tides by means of harmonic constants, International Hydrographic Bureau, Monaco, 1926.

18. Cesare Marchetti, private communication.

19. I. Browning, *Climate and the Affairs of Men,* (Burlington, VT: Frases, 1975).

20. See Gerald S. Hawkins in note 16 above.

Chapter Nine: Reaching the Ceiling Everywhere

1. Cesare Marchetti, "The Automobile in a System Context: The Past 80 Years and the Next 20 Years," *Technological Forecasting and Social Change,* vol. 23 (1983): 3–23.

2. The data on subway inaugurations used for the determination of these curves come from: Dominique and Mechèle Frémy, QUID 1991 (Paris: Robert Lafont, 1991), p. 1634.

3. Theodore Modis and Alain Debecker, "Innovation in the Computer Industry," *Technological Forecasting and Social Change,* vol. 33, (1988): 267–78.

4. The airway curve mentioned here is calculated by me and shown in Figure 9.5.

5. Michael Royston, private communication.

6. This ceiling has been estimated in Cesare Marchetti, "Fifty-Year Pulsation in Human Affairs, Analysis of Some Physical Indicators," *Futures,* vol. 17, no. 3, (1986): 376–88.

7. This argument has been made by Cesare Marchetti in "On Transport in Europe: The Last 50 Years and the Next 20," invited paper, First Forum on Future European Transport, Munich, September 14–16, 1987.

8. The name Maglevs and their role as presented here have been suggested by Cesare Marchetti; see note 7 above.

9. Larry Abraham's *Insider Report,* vol. VII, no. 9, 1990.

10. John Doe's [pseudonym] *Report from Iron Mountain on the Possibility and Desirability of Peace* (New York: The Dial Press, 1967).

11. *Historical Statistics of the United States, Colonial Times to 1970,* vols. 1 and 2, Bureau of the Census, Washington, DC, and the *Statistical Abstract of the United States,* U.S. Department of Commerce, Bureau of the Census.

12. Yacov Zahavi, Martin J. Beckmann, and Thomas F. Golob, *The Unified Mechanism of Travel (UMOT)/Urban Interactions* (Washington, DC: U.S. Department of Transportation, 1981).

13. The data source is the same as in note 11 above.

14. John Naisbitt, *Megatrends,* (New York: Warner Books, 1982).

15. See note 7 above.

Chapter Ten: If I Can, I Want

1. P. Ouspensky, *In Search of the Miraculous,* (New York: Harcourt, Brace & World, 1949).

2. For a brief description see Appendix A. For more details see reference therein and ultimately Vito Volterra, *Lecons sur la théorie mathématique de la Lutte pour la vie,* (Paris: Jacques Gabay, 1931).

3. D. A. MacLuich, "University of Toronto Studies," *Biological Science,* vol. 43 (1937).

4. Elliott W. Montroll and Wade W. Badger, *Introduction to Quantitative Aspects of Social Phenomena* (New York: Gordon and Breach Science Publishers, 1974), pp. 33–34.

5. See note 4 above, pp. 129–30.

6. H. O. Peitgen and P. H. Richter, *The Beauty of Fractals* (Berlin and Heidelberg: Springer-Verlag, 1986).

7. P. F. Verhulst, "Recherches mathématiques sur la loi d'accroissement de la population," *Nouveaux Mémoires de l'Académie Royale des Sciences et des Belles-Lettres de Bruxelles,* vol. 18, (1845): 1–40.

8. James Gleick, *Chaos,* (New York: Viking, 1988).

9. See note 4 above, p. 33.

10. The computer simulation was carried out by Z. Fortune, as cited in Cesare Marchetti, "The Automobile in a System Context: The Past 80 Years and the Next 20 Years," *Technological Forecasting and Social Change,* vol. 23 (1983):3–23.

Chapter Eleven: Forecasting Destiny

1. The remaining of this chapter is based on Cesare Marchetti's ideas about energy, society, and free will. It draws from two of his works: "Energy Systems—the Broader Context," *Technological Forecasting and Social Change,* vol. 14, (1989): 191–203, and "On Society and Nuclear Energy," report EUR 12675 EN, 1990, Commission of the European Communities, Luxemburg.

2. The first example on U.K. coal production was mentioned by Cesare Marchetti in his article, "Fifty-Year Pulsation in Human Affairs, Analysis of Some Physical Indicators," *Futures,* vol. 17, no. 3, (1986): 376–88. The second example was mentioned in "Energy Systems—the Broader Context"; see note 1 above.

3. A different graph with partial data was originally used by Cesare Marchetti to illustrate these ideas in the article entitled "Energy Systems—The Broader Context"; see note 1 above.

4. Cesare Marchetti, "On Society and Nuclear Energy"; see note 1 above.

5. A. Mazur, "The Journalists and Technology: Reporting about Love Canal and Three Mile Island," *Minerva,* vol. 22, no. 1: 45–66, as cited by Cesare Marchetti, "On Society and Nuclear Energy"; see note 1 above.

6. These arguments have been put forward by Cesare Marchetti in "On Society and Nuclear Energy"; see note 1 above.

7. J. Weingart, "The Helios Strategy," *Technological Forecasting and Social Change,* vol. 12, no. 4 (1978).

8. This section is based on Cesare Marchetti's article, "Energy Systems—the Broader Context"; see note 1 above.

9. This image appears in Cesare Marchetti, "Branching Out into the Universe," in Nebojsa Nakicenovic and Arnulf Grubler, eds., *Diffusion of Technologies and Social Behavior* (Berlin: Springer-Verlag, 1991).

10. J. Virirakis, "Population Density as the Determinant of Resident's Use of Local Centers. A Dynamic Model Based on Minimization of Energy,"

Ekistics, vol. 187, (1971): 386, as cited by Cesare Marchetti in "Energy Systems—the Broader Context"; see note 8 above.

Appendix A: Mathematical Formulation of S-Curves and the Procedure for Fitting Them on Data

1. E. W. Montroll and N. S. Goel, "On the Volterra and Other Nonlinear Models of Interacting Populations," *Review of Modern Physics,* vol. 43, no. 2 (1971): 231.
2. M. Peschel and W. Mendel, *Leben wir in einer Volterra Welt?* (Berlin: Akademie Verlag, 1983).
3. See note 1 above.
4. P. F. Verhulst, *Nouveaux Mémoires de l'Académie Royale des Sciences, des Lettres et des Beaux-Arts de Belgique,* vol. 18 (1845): 1–38.
5. J. B. S. Haldane, "The Mathematical Theory of Natural and Artificial Selection," *Transactions, Cambridge Philosophical Society,* vol. 23 (1924): 19–41.
6. Alfred J. Lotka, *Elements of Physical Biology* (Baltimore: Williams & Wilkins Co., 1925).
7. Nebojsa Nakicenovic, "Software Package for the Logistic Substitution Model," report RR-79-12, 1979, International Institute for Applied Systems Analysis, Laxenburg, Austria.

Appendix B: Expected Uncertainties on S-Curve Forecasts

1. Alain Debecker and Theodore Modis, "Uncertainties in S-curve Logistic Fits," Presented at the International Conference on Diffusion of Technologies and Social Behavior, Laxenburg, Austria, June 14–16, 1989; also at the 6th International Symposium of Forecasters, Paris, France, May 1986.

ACKNOWLEDGMENTS

I am gratefully indebted to the work of various scientists in Western Europe and the United States during the last 150 years. These people have shared a passion for the same subject that now fascinates me, and their contributions to this work are cited in the Notes and Sources. Among them, Cesare Marchetti requires special mention. He has contributed enormously to making the formulation of natural growth a general vehicle for understanding society. I draw extensively on his ideas, ranging from the concept of invariants (Chapter One) to the productivity of Mozart (Chapter Four) and the evolution of nuclear energy (Chapter Seven). I present many of his results, and I know that he has inspired many of the results obtained by others. He has masterminded the explanation of the intricate interdependence of primary energy systems and transport infrastructures discussed in Chapter Nine. The culminating Chapter Eleven is largely based on two of his works, "Energy Systems" and "On Society and Nuclear Energy." Thus, this book could not have come into existence had it not been for the work of Cesare Marchetti and of his collaborators, Nebojsa Nakicenovic and Arnulf Grubler, at the International Institute for Applied Systems Analysis in Laxenburg, Austria. I am grateful to them for keeping me abreast of their latest work and for all the fruitful discussions we have had together.

Of great importance has been the contribution of my collaborator, Alain Debecker. His personal and professional support reflect on many aspects of this book. I also want to thank another colleague, Phil Bagwell, my initiator in the world of business, for his continuous encouragement and guidance.

The idea for this book was given to me by Michael Royston, who encouraged and advised me through the phases of conception. Gordon Frazer helped my writing when he revealed to me—in a condensed tutorial session on composition—how much one needs not to say. The text further benefited from the editorial work of Marilyn Gildea, who reviewed the English with an exemplary professionalism, teaching me much about how to write correctly. She also won my confidence on suggestions well beyond the editorial scope.

My agent, John Ware, has demonstrated why authors need agents. He has also become my friend. My editor at Simon & Schuster, Fred Hills, acted as my mentor: severe, supportive, farsighted, and wise. His colleague, Burton Beals, worked hard on the manuscript and made me do the same, thus producing a greatly improved text.

I am also indebted to Pierre Darriulat, Simon van der Meer, John Casti, Nebojsa Nakicenovic, and Nikos Nikas who were kind enough to read the manuscript and offer their comments on short notice. There are a number of others who directly or indirectly helped the completion of this work: Dimitri Peretzis, Marc van der Erve, Ziba Khalat-Bari, Eva Luraschi, and Mireille Forestier.

I want to thank two special people, Michel de Saltzmann and Mihali Yannopoulos, my teachers in some form or another, something that has influenced several aspects of this book.

Finally, I want to thank my children, Yorgo and Thea, who empathized with me as I went through the growing pains of writing a book during the last four years.

Theodore Modis
Geneva, Switzerland
March 1992

INDEX

Page numbers in *italics* refer to figures.

abortion, 31
Abraham, Larry, 182
accidents, 82–83
  airplane, 213, *272–73*
  car, 24–26, *25,* 98, 142, 207, *213,*
    272
  nuclear, 211–14, *212*
Action Curves and Clockwork Geniuses
  (Marchetti), 79–80
Afghanistan, Soviet intervention in,
  122
age:
  at death, 30–31
  of Nobel laureates, 94
  "old," Red Brigades and, 97
AIDS, 104–5, 115, 144, *254, 264*
"AIDS Surveillance Report, The,"
  105
air traffic:
  total worldwide, 167, 169, *174, 268*
  passenger worldwide, *174*
air travel:
  competition and, 137, 138, *168,*
    263

  echo effect and, 213, *272–73*
  saturation and, 167, 178–80, *268*
  between U.S. cities, 171, 180
aircraft (passenger) performance,
  173
alcoholism, 158
"America's Longest War" (McNeil),
  116
ammonia production, efficiency of,
  216, *274*
animals:
  age at death of, 30–31
  conservation vs. innovation of,
    125
  work, 152
antibiotics, 103
anti-Semitism, 162
Aponi (hypothetical machine), 28–30
archaeology, S-curves and, 51–52
Armstrong, Neil A., 193
Asimov, Isaac, 30
astrology, 19–20
Athens, 219–20
Athens Institute of Ekistics, 219–20

287

Aubrey holes, 165
Ausubel, J., 203
automobiles, *see* cars
Ayres, R., 156
Aztecs, 182

Bach, Johann Sebastian, 39–40, 193
backcasting:
    Christianity and, 72
    explorations and, 51
bacterial growth, 59, *240*
bank failures, *150,* 151
Bartók, Béla, 223
Beauty of Fractals, The (Peitgen and
    Richter), 196, *197*
bell-shaped curves:
    of life cycle, 32–34, *33, 34,* 38–
        40, *237*
    normal (Gaussian), 38–40, *237*
Bible, 31
birth rate, 85–86, *248*
body language, 119
Brahms, Johannes, 77, 82, *245*
Brookhaven Natural Laboratory, 65
BULL, 88
Bureau of Labor Statistics, U.S., 114
business:
    collective learning and, 44–49
    competition in, 45–46
    growth of, 35–36
    Time Series approach and, 21
    see also companies; products

Cadillac Cyclone, *168*
calendars:
    Chinese, 164–65
    Middle American, 182
Canada, cars in, 171
canals, 178, *178,* 179, *179*
    competition and, 136, 137, 138,
        263
cancer, 98, 99–100, 103, 115, 143–
    144, *264*

canonization, rate of, 70–72, *71*
cardiovascular disease, 98, 115, 143–
    144, *264*
cars, 114, 137, 140, 171
    accidents and, 24–26, *25,* 98, 142,
        207, *213, 272*
    future of, 185–87
    growth in number of, 56–57, 154,
        239
    saturation and, 154–55, 166–67,
        168, 169
    substitution and, 107–10, *108, 110*
catalytic converters, *174*
"catching-up" effect, 203, *203*
    canonization and, 71
    Hitchcock and, 132
cathedrals, Gothic, 64–65, 66, 243
chain letters, breakdown of, 60–61
chaos, 44, 154, *197*
    coal and, 200–201, *201,* 203
    defined, 196
    energy consumption and, 203–4,
        271
    Japanese new car registrations and,
        198, *199*
    plywood sales and, 199, *270*
    from S-curves to, 195–204, *197,*
        199, 201, 202
Chaos (Gleick), 20, 44
charge phase, 175, *175*
chemical elements, stable:
    clustering of discoveries of, 129,
        130, 132
    energy consumption and, *150,* 151
    learning and, 44
chess, 42
China Syndrome (film), 213
Chinese calendar, 164–65
Chopin, Frédéric, 223
Christianity, canonization rate and,
    70–72, *71*
cirrhosis of the liver, 143, 158, *159*
Club of Rome, 13–14

clustering, 127–33
 of discoveries of stable chemical
  elements, 129, *130,* 132
 of innovations, 127–29, *128,* 171
 niche-within-a-niche and, 131–32,
  131
coal, 69, 152, 171, 215
 chaos and, 200–201, *201,* 203
 competition and, 139, *139,* 140,
  142
 in Great Britain, 208, *209*
coffee machines, 205–6, 207
collective learning, 43–54
 of companies, 44–49
 exploration and, 49–53, *51*
 of scientific community, 44
Columbus, Christopher, 49–51, *51,*
  191, 193
communications, 118–19, 186, *260*
companies:
 collective learning of, 44–49
 conservation vs. innovation of,
  125
 criminal organizations compared
  with, 97
 defined, 97
 see also specific companies
competition, 91–105, 124–46
 in business, 45–46
 clustering and, 127–33, *128, 130,*
  131
 creativity and, 73
 criminal activities and, 95–97
 discoveries and, 191
 disease and, 97–105, 143–44, *264*
 between energy sources, 138–43,
  139
 learning and, 42–43, 45–46, 49
 in logistic growth, 35, 36
 Marchetti's views on, 12, 14
 natural selection and, 124–25
 Nobel prizes and, 92, *93,* 94, 144–
  146, *265*

in Olympic Games, 135–36
saturation and, 134–35, 138
social relationships and, 171, 174
substitutions and, 106–23, 133–46,
  134; see also substitutions
 between transport infrastructures,
  136–38, *168, 263*
 U.S. vs. foreign, 94, 95, 145
 war as, 124
computer programmers, 27–28
computers (computer innovation),
  14, 15, *174*
 creativity/productivity and, 87–89,
  251
 Digital, 46–48, 111; *see also* VAX
  family products
 efficiency and, 215–17
 MIPS and, 46–48
 population growth of, 57, *58,* 59
 rationing of, 26–27
 software for, 27
Concorde, 137
Cosmotron, 65
creativity/productivity, 73–91
 of Brahms, 77, 82, *245*
 computer innovation and, 87–89,
  251
 death predictions and, 79–81
 of Einstein, 78–79, *78*
 failures of S-curve approach to,
  81–85
 of Fellini, 86–87, *249*
 "fertility" template and, 85–86,
  248
 of Hemingway, 82, *246*
 of Hitchcock, 131–132, *131*
 Marchetti and, 73–74, 79–80, 83
 of Marquez, 86–87, *249*
 of Mozart, 74–77, 82
 of Richter, 86–87, *249*
 of Rubbia, 86–87, *249*
 of Schumann, 83–85, *84,* 89
 of Shelley, 82–83, *247*

crime, criminals:
    careers of, 95–98, 252
    fifty-six-year cycle and, 160, *161*
"culture performance driver," 126
cumulative growth, 33–35, *34*
    see also S-curves
Czechoslovakia, Soviet intervention
    in, 122

D'Ancona, Umberto, 11
Darwin, Charles, 124
Dead Sea scrolls, 72
death:
    age at, 30–31
    from car accidents, 24–26, *25,* 98,
        142, 207, 213, *272*
    disease and, 97–105, *99, 102, 253,*
        254
    echo effect and, 213, *272–73*
    prediction of, 79–81
    rate of, 98–99, *99*
    unnatural, 82–83
Debecker, Alain, 26, 37, 87–88, 169
    chaos and, 200–201
    uncertainties and, 234–35, *235–36*
decision making:
    events contrary to, 208–14
    optimization and, 221
decline, observational approach to,
    92
de Moivre, Abraham, 38*n*
Depression, Great, 114
detergents, substitution and, 112,
    113, 114, 119
diffusion model, 35
Digital Equipment Corporation
    (DEC), 27, 46–48, 111, 126
    see also VAX family products
diphtheria, 100–104, *102*
discharge phase, 175, *175*
discoveries, 191
    clustering of, 127, 129, *130*
    random, 41–44

of stable elements, 44, 129, *130,*
    132, *150,* 151
diseases:
    competition between, 97–105,
        143–44, *264*
    substitution and, 122, 143–44,
        264
    see also specific diseases
Dissonant Quartet in C major K465
    (Mozart), 77
drosophila (fruit flies), population
    growth of, 56–57, *239*

Earth Day, 185
echo effect, 213, *272–73*
eclipses, 163
economic forecasting, 20–21
economic growth, 175, 181–82
    energy consumption and, 151,
        152, 154
economies of scale, 48, 70
efficiency, 215–18, *274*
Einstein, Albert, 32, 78–79, *78*
electronics industry, 125
Elements of Physical Biology (Lotka),
    62–63
energy, 23, 67–70
    competition between sources of,
        138–43, *139*
    economics of scale and, 70
    forecasting of demand for, 12
    money transfers compared with,
        23
    prices of, 152, *153,* 157, 215
    renewable, 114, 139, 214, *256*
    substitution of, 69–70, 114, 139–
        141, 152, *152–53,* 208–14, *209,*
        210, 212, 217–18, 256
    see also coal; natural gas; nuclear
        energy; oil; wood
energy consumption, 147–55, *148*
    activities that echo cycle of, 149,
        150, 151, 157–62, *159, 161*

chaos and, 203–4, *271*
efficiency and, 217
substitution and, 152, *152–53*
engines:
  efficiency of, 216, *274*
  internal-combustion, 133, *134*
  steam, 109, *255*
environment, 184
  see also pollution
environmentalists, 125, 141–42
equal-temper tuning, 39
equilibrium:
  homeostasis, 24
  see also invariants
ergodic theorem, 188–92
Essenes, 72
European Center for Nuclear Research (CERN), 65
evolution, natural selection and, 124–
  125, 224
exploration, 49–54
  Columbus and, 49–51, *51*
  of moon, 52, *238*
  tourism compared with, 191, 193,
  194
exponential growth, 35, 37, 147

farming machines, 116, 118, *259*
feedback loops, 163, 224
  defined, 68*n*
  media-public, 212–13
  oil and, 68
Fellini, Federico, 86–87, *87, 250*
"fertility" template, 85–86
fifty-six-year cycles, 156–65, 181–82,
  225
  bank failures and, *150,* 151
  cirrhosis of the liver and, 158, *159*
  eclipses and, 163
  energy consumption and, 148–51,
  148, 150
  energy prices and, 152, *153*
  explanations of, 162–65

homicides and, 160, *161*
horsepower use and, 149, *150*
innovation appearance and 149–
  150, *150*
life expectancy and, 157–58, *159,*
  160
murder weapon and 160,
  161
one-mile run and, 158, *159,* 160
primary energy substitution and,
  152, *153*
Royston spiral and, 174–77, *175*
stable elements discovery and, 151,
  150
stock market crash of 1987 and,
  156–57
Stonehenge and, 165
subway network construction and,
  151
sunspot activity and, 164, *267*
U.K. Wholesale Price Index and,
  176, *177*
women Nobel laureates and, 159,
  160
film:
  creativity/productivity and, 86–87,
  87, 250
  of Hitchcock, 131–32, *131*
Fisher, J. C., 35, 106–7, 112, 119–
  120, 122–23
fish populations, growth rate of,
  11–12
fitness:
  success and, 45
  survival and, 36, 45, 107, 124–25
fluoridation, 116
forecasting:
  destiny, 205–21
  economic, 20–21
  of energy demand, 12, 214, 217–
  218
  of energy sources, 140, 141–143
  of invariants, 23–32

forecasting (*cont.*)
  physics and, 22–23
  Time Series, 21
  weather, 20
  see also S-curves
foretelling, science and, 19–40
Formula One racing, 220–21
Foster, Richard, 126
France, telephone calls vs. letters in,
  118–19, *260*
free will, 16, 218–21
fruit flies (drosophila), population
  growth of, 56–57, *239*
futuronics, 181–87
  cars and, 185–87
  city sizes and, 185–187
  war and peace and, 182–85

Galton, Francis, 22*n*–23*n*
Garcia Marquez, Gabriel, 86–87,
  87, 249
gas:
  natural, 67, 139, *139*, 141–42
  synthetic, 141
Gauss, Karl Friedrich, 38–40
Gaussian (normal) curve, life
  cycle compared with, 38–40,
  237
genetic predetermination, 61–62
Gleick, James, 20, 44, 196, 198
Gleick's Butterfly Effect, 22
global village, 185–87
Goel, N. S., 229, 230
gold, 67–68
Gothic cathedrals, 64–65, 66, *243*
Grand Unification, 32
gravity, 32
Great Britain:
  energy in, 208, *209*
  opposition to technological inno-
    vations in, 116–118, *259*
  Wholesale Price Index of, 176–77,
    177

growth:
  cumulative, 33–35, *34; see also*
    S-curves
  rate of, 32–33, *33, 34; see also* life
    cycles
growth functions, 35–36
Grubler, Arnulf, 109, 118–19, 137,
  169, *172,* 203

Haldane, J. B. S., 231
hares, oscillations of, 195
harpsichords, tuning of, 39
Hawkins, Gerald, 165
haystacks, finding needles in, 41,
  91
Health Department, U.S., 105
heartbeats, number of, as invariant,
  30–31
heating system, 68*n*
height:
  bell curves and, 32, *33, 34*
  prediction of, 61–62
  S-curve and, 33–34, *34,* 35
Heisenberg, Werner, 215
Hemingway, Ernest, 82, *246*
Heraclitus, 124
highways, *see* roads and highways
Historical Statistics of the United States,
  136, 147, 149
Hitchcock, Alfred, 131–32, *131*
Ho, 19
Hobsbawm, E. J., 118
homeostasis, 24
homicides, 160, *161*
  echo effect and, 213, *272–73*
Honeywell, 88
horoscopes, 19–20
horsepower, 149, *150*
horses, substitution and, 107–10, *108,*
  110
Hsi, 19
Huang Ti, Emperor of China,
  164–65

Hungary, Soviet intervention in, 120, 122

IBM computers, 27, 89, 111
income, allocation and, 15, 185
industrial growth, 181–82
    energy consumption and, 151
    timing of, 169, 171, *172–74*
industry learning curve, 46–49
    computers and, 46–48
infants:
    mortality of, 30, 31
    vocabulary acquisition of, 43, *43,* 130–31
information exchange, 186
information workers, 114–15, *257*
innovation:
    clustering of, 127–29, *128,* 171
    cultural resistance to, 116, *117,* 118, *259*
    economic growth and, 152, 154
    energy consumption and, 149, *150,* 151
    saturation and, 155, 171
    survival and, 125–26
    see also technological change
In Our Time (Hemingway), 82
In Search of the Miraculous (Ouspensky), 39–40
Insider Report, 182
International Institute of Advanced Systems Analysis (IIASA), 12–15, 169
International Conference on Diffusion of Technologies and Social Behavior (1989), 83
invariants, 14–15, 23–32, 181, 207
    car safety and, 24–26, *25*
    computer innovation and, 26–27
    defined, 12
    futuronics and, 185–87
    humans vs. machines and, 27–30

number of heartbeats as, 30–31
    turning into variables, 31–32
Italy, cars in, 56–57, 154, *239*

Jackson, Henry M., 137
Japan:
    cars in, 154, 166–67, 198, *199*
    Maglevs in, 180
    substitution of detergent for soap in, 112, *113,* 114
Jesus, 72
jet engine performance, *173*
Jevons, William S., 155–56

Kleinknecht, Alfred, 156
Koechel, Ludwig, 74–75
Kondratieff, N. D., 156, 177
Kondratieff barrier, 177, 178
Kondratieff cycle, 156, 181, 182
Kondratieff wave, 177

La Fontaine, Jean de, 218
Lawrence, George H., 141
learning, learning curves, 41–54
    archaeology and, 51–52
    economics of scale and, 48
    exploration and, 49–54, *51, 238*
    industry, 46–49
    life cycle of, 41–42
    repetition and, 42
    of scientific community, 44
    from success, 44–49
    vocabulary acquisition and, 43, *43*
    see also collective learning
Lenin, 185
LEP (electron accelerator), 65
letters, substitution and, 118–19, *260*
Lewin, Leonard C., 182–84
life cycles, 17, 32–40
    bell-shaped curve of, 32–34, *33, 34*
    calculated from S-curve, 34–35
    defined, 32

life cycles (*cont.*)
  discovery of stable elements and, 129–30, *130*
  of learning process, 41–42
  of musician, 34
  normal (Gaussian) distribution compared with, 38–40, *237*
  optimism vs. pessimism and, 91–92
  product, 34
  saturation phase and, 135
  S-curve jumping and, 126, *127*
  S-curve obtained from, 33
life expectancy, 30–31
  fifty-six-year cycle and, 157–58, *159*, 160
light production, efficiency of, 216, *274*
Limits to Growth, The, 13–14
liver, cirrhosis of, 143, 158, *159*
living conditions, improvement of, 98, 103
logistic, 35–40, 47, 124, 126, 188, 229–33
  chaos and, 44
  as diffusion model, 35
  normal distribution compared with, 38–40, *237*
  performance/price indicator and, 48
  straight lines and, 111–12, *113*
London, subway in, 151
longevity, 225–26
Lotka, Alfred J., 12, 55–56, 59, 124, 231
  railway networks and, 62–63, *241*
  see also Volterra-Lotka system of differential equations
lunar eclipses, 163
lynx, oscillations of, 195

McKinsey & Company, 126
McLuhan, Marshall, 186
McNeil, Donald R., 116

Maglev (magnetic levitation train), 178, 180, 186
Makridakis, Spyros, 21
manual laborers, substitution and, 115, *257*
Marchetti, Cesare, 12–16, 23, 123, 162, 167, 169, 187, 210
  background of, 13
  on canonization rate, 70, 72
  creativity/productivity and, 73–74, 79–80, 83
  on criminal careers, 95–96
  energy and, 138–40, *139*, 152, 154, 211–13, *212*, 218
  innovation and, 129
  life expectancy and, 31
  longevity of, 225–26
market niche, 15, 36, 38, 59, 111
  of cars, 154
marketers, 218–19
marriage, length of, 171
Matterhorn, expeditions to, 52–53
Mayans, 182
medicine, 98, 103
Megamistakes (Schnaars), 107
Megatrends (Naisbitt), 186
Mendel, W., 229
Mensch, Gerhard, 128–29
merchant vessels, substitutions and, 133, *134*
Mesopotamia, archaeology in, 51–52
methanol, 141
Meton, cycle of, 163
microniche, 67, 111, 121
MicroVAX II, 57, 59
money, transfers of, 23
Monte Carlo study, 234–35, *235–36*
Montrey, Henry, 199
Montroll, Elliott, 23, 198, 229, 230
moon explorations, 52, *238*
mountain climbing, 52–53
movement, invariants and, 14–15
movies, *see* film

Mozart, Wolfgang Amadeus, 74–77,
  82, 95
murder weapons, 160, *161*
musicians, 223–24
  creativity/productivity of, 74–77,
    83–85, *84,* 89, *245*
  cumulative performances of, 34
  life cycles of, 32–33, 34
  technical virtuosity of, 28–30
mutations, 125, 224

Nader, Ralph, 26, 142
Naisbitt, John, 186
Nakicenovic, Nebojsa, 114, 134, 138,
    152, 169, 178, 203, 214
  multiple competition and, 232
  Wholesale Price Index and, 176–
    177, *177*
National Climate Analysis Center,
    164
natural gas, 67, 139, *139,* 141–42
  pipelines of, 171, *173*
natural selection, 124–25, 224
new impressions, value of, 192–93
Newton, Sir Isaac, 32
New York, N.Y., subway in, 151
New Zealand, cars in, 171
niche, 14, 36–38, 67–68, 81, 99,
    111, 121, 224, 231
  of AIDS, 105, 144
  of beauty soap, 114
  of bituminous coal, 200
  capacity of, 231
  of cars in society, 56, 154, 166,
    198, 199
  of computer manufacturers, 88
  of computer models, 88
  definition of, 111, 207
  of diseases, 99
  ecological, 56, 64, 98, 196
  emptying of, 67, 91, 100, 104
  of executives, 116
  for generalists in the future, 181

of home vocabulary, 43, *43,* 130
of imported oil, 210
of inconvenience, 207
of innovations, 154
invariants and ceiling of, 207
of knowledge, 52
of Nobel prizes, 92
of nuclear energy, 142, 211
overtaking of, 154
of plywood, 199, *270*
of railways, 63
rate of growth correlated with size
    of, 235
of roads, 154
size of, 73
of transport, 108
niche-beyond-a-niche, 119, 203, *271*
niche-within-a-niche, 81, 89, 119,
    131, *131,* 132
  clustering and, 131–32, *131*
  substitutions and, 119
Nixon administration, 137
Nobel prizes:
  of Americans, 92, *93,* 94, 145
  competition for, 92, *93,* 94, 144–
    146, *265*
  of Swedes vs. Russians, 121–22,
    262
  women and, *159,* 160
normal (Gaussian) curve, life cycle
    compared with, 38–40, *237*
nuclear energy, 67, 139, *139,* 142–
    143
  accidents and, 211–14, *212*

octave, natural, 39–40
oil, 67–69, *69, 139,* 140–41, 142,
    152, 171, 215
  car growth and, 57
  pipeline construction of, 171, *173,*
    269
  U.S. imports of, 208–10, *210*
Olympic Games, 135–36

one-mile run, fifty-six-year cycle and, 158, *159,* 160

"On Society and Nuclear Energy" (Marchetti), 211–13, *212*

"On 10^{12}" (Marchetti), 14

optimism, 91–92

optimization, free will and, 220–21

order:
  chaos and, 44
  in random discoveries, 41–44

oscillations, 195–96, *197,* 198, 207, 224

Ouspensky, Peter, 39–40, 192

pandas, 125

particle accelerators, 64, 65–66, *244*

patristic curve, 70, *71,* 72

peace, possibility and desirability of, 182–85

Peitgen, H. O., 196, *197*

perceived target, 73

performance/price indicator, 46–49

Peschel, M., 229

pessimism, 91–92

Petroff, S. A., 103

phasing out, 57, 138

Phillips, David P., 213

physics:
  ergodic theorem and, 188–92
  forecasting and, 22–23
  invariants in, 31–32

pianos, tuning of, 39

plywood sales, 199, *270*

police operations, effectiveness of, 96

pollution, 107–8, 154, 171
  war on, 183–85

population curves, population growth, 55–67
  bacterial growth and, 59, *240*
  canonization rates and, 70–72, *71*
  car growth and, 56–57, *239*
  of chain letters, 60–61
  of computers, 57, *58,* 59

energy sources and, 67–70, *69*
  of fish, 11–12
  of fruit flies, 56–57, *239*
  gold and, 67–68
  of Gothic cathedrals, 64–65, 66, *243*
  inanimate, 35–36, 56–67, *58*
  natural selection and, 124–25
  oil and, 67–69, *69*
  oscillations and, 195–96
  particle accelerators and, 64, 65–66, *244*
  of rabbits, 36–38, 56
  railway networks and, 62–63, *241*
  supertankers and, 63–64, 66, *242*
  Volterra-Lotka equations and, 55–56

predator-prey equations, 195, 229–30

Price, Derek J. de Solla, 22

prices:
  of energy, 152, *153,* 157, 215
  of products, 45–49

prison confinement, effects of, 96

productivity, *see* creativity/ productivity

products:
  collective learning and, 44–49
  costs of, 45, 48
  cumulative number sold of, 34
  customer needs and, 45
  Digital VAX family, 46–48, 57, *58, 59*
  growth function of, 35–36
  life cycle of, 34
  phasing out of, 57
  price of, 45–59
  technological component of, 44–45

Pry, R. H., 35, 106–7, 112, 119–20, 122–23

punishment, 96

rabbit populations, growth of, 36–38, 56

railways, 109, 178–80, *178*
  competition and, 136–38, *168, 263*
  fuel for, 140
  future of, 186
  growth of, 62–63, *241*
  Maglev and, 178, 180, 186

random discoveries, order in, 41–44

rats, oscillations of, 195

recession, 157, 175, 181
  saturation and, 154–55, *170,* 171, 177

Red Brigades, 97, 98, *252*

relationships, cycle of, 171

relativity theory, 32, 118

relaxation phase, 175, *175*

renewal rituals, 182

repetition, learning and, 42

Report from Iron Mountain on the Possibility and Desirability of Peace (Lewin), 182–84

Richter, Burton, 86–87, *87, 249*

Richter, P. H., 196, *197*

Rise and Fall of Infrastructures, The (Grubler), 109, 169, *172*

roads and highways, 109, 154, 167, 171, 178–80, *178, 179*
  competition and, 137, 138, *263*

Rosehooded, Tim, 190–91

Royston, Michael, 174–77

Royston spiral, 174–77, *175*

rubber, natural vs. synthetic, 120, 210–11, *261*

Rubbia, Carlo, 86–87, *87, 250*

Rude, G., 118

Rutherford, Ernest, 66

safety, car, 24–26, *25,* 142

saints, canonization of, 70–72, *71*

Sanford, T. W. L., 65–66

Saros, cycle of, 163

saturating competitor, 134–35

saturation, 166–76
  air traffic and, 167, *268*
  cars and, 154–55, 166–67, *168,* 169
  recession and, 154–55, 169, *170, 171,* 177
  S-curves and, 169, *170,* 171
  transportation and, 177–81, *178, 179*

Schnaars, Steven, 107

Schumann, Robert, 83–85, *84,* 89

Schumpeter, Joseph A., 156

scientific method, 16, 20, 224
  creativity/productivity of, 78–79, *78,* 86–87, *87, 249, 250*
  Marchetti's use of, 12

S-curves, 17, 33–40, 181
  chaos and, 195–204, *197, 199, 201, 202*
  clustering and, 129–32, *130, 131*
  competitive substitutions and, *see* substitutions
  creativity and, *see* creativity/productivity
  criminal careers and, 95–97, *252*
  of cumulative growth, 33–35, *34*
  discrete, *202–3*
  diseases and, 97–105, *99, 102, 253, 254*
  fitting of, 37, 232–33
  jumping, 126, *127*
  learning process and, *see* learning, learning curves
  life cycle calculated from, 34–35
  logistic and, 35–40
  mathematical formulation of, 35, 229–33
  for Nobel prizes, 92, *93,* 94
  obtained from life cycle, 33
  population growth and, *see* population curves, population growth
  saturation and, 169, *170,* 171

298 INDEX

S-curves (*cont.*)
   symmetry of, 59–60
   tourism and, 53–54, 193–95
   uncertainties and, 37–38, 51,
      234–36
   U.S. energy consumption and,
      147–48, *148, 266*
   volume curve and, 48–49
Senate, U.S., 137
sexual discrimination, homicides and,
      160, *161*
sharks, 125
sheep, oscillations of, 196
Shelley, Percy Bysshe, 82–83, *247*
ships, 133, *134*
smoking, 118
soap, substitution and, 112, *113,* 114,
      119
social system, innovation in, 129
software, computer, 27
solar eclipses, 163
solar energy, *139,* 214
solfus, *139*
Soviet Union:
   foreign interventions of, 120, 122
   Nobel prizes and, 121–22, *262*
   road construction in, 171
   steam engine in, 109, *255*
Spanish flu, 104
Stalemate in Technology (Mensch),
      128–29, *128*
Statistical Abstract of the United States,
      101
steady state, 196, *197*
steam engines, substitutions and, 109,
      173, 255
steamships, 133, *134*
steel industry, 125
Steiner, Rudolf, 111
Stewart, Hugh B., 149
stock market crash (1987), 155–57
Stonehenge, 163, 165
Stonehenge Decoded (Hawkins), 165

straight lines, substitutions and, 111–
      116, *113, 133–34, 134,* 136,
      256–58
strikes, 140, 208, *209*
substitutions, 106–23, 133–46, *134*
   of cars for horses, 107–9, *108*
   cultural resistance and, 116, *117,*
      118, *259*
   of detergent for soap, 112, *113,*
      114, 119
   of energy sources, 69–70, 114,
      139–41, 152, *152–53,* 208–14,
      209, 210, 212, 217–18, *256*
   epidemics and, 122
   of information workers for manual
      laborers, 114–15, *257*
   of margarine for butter, *261*
   Nobel prizes and, 121–22, *262*
   steam engines and, 109, *255*
   straight line character of, 111–16,
      113, 133–34, *134,* 136, *256–58*
   of Swedes for Russians, 120–122,
      262
   successive, 133–35, *134*
   of synthetic for natural fibers, *261*
   of synthetic for natural rubber,
      120, *261*
   of telephone calls for letters, 118–
      119, *260*
   unnatural, 118–20
   of women executives for men ex-
      ecutives, 115–16, *258*
subways, 151, 180–81, *172, 174*
success:
   competition and, 45–46
   learning from, 44–49
suicide:
   echo effect and, 213, *272*
   of Hemingway, 82
   Schumann's attempting of, 83–85
sunspot activity, 164, *267*
Superconducting Super Collider
      (SSC), 65

supersonic transport, 137, 179, 180, 186
supertankers, 63–64, 66, *242*
survival:
    conservation vs. innovation in, 125–26
    of the fittest, 36, 45, 107, 124–25
    of Nobel prizes, 121–22, *262*
symmetry, 81
    of S-curves, 59–60

technological change:
    diffusion of, 106–7
    resistance to, 116, *117*, 118, *259*
telegrams, 118, 119
telephone calls, substitution and, 118–19, *260*
television, Hitchcock's work for, *131*, 132
telexes, 119
Tell, William, 46
tension phase, 175, *175*
terrorism, 97, 98, *252*
thermodynamics, 22, 23
thermonuclear fusion, *139*
Thomas Aquinas, Saint, 70, *71*
Thomistic curve, 70, *71*
Thornton, H. G., 59
Three Mile Island, *212*, 213
tic-tac-toe, 42
Time Series, 21
Timing of Technological Growth Curves, The (table), 169, *172–74*
tourism, 191–95
    S-curves and, 53–54, 193–95
trains, *see* railways
transportation:
    competition and, 136–38, *263*
    future of, 185–87
    saturation and, 177–81, *178*, *179*
    ships, 133, *134*

subways, 151, 180–81
    see also air travel; cars; railways
travel time, 14–15, 185
tuberculosis, 98, 103–4, *253*
tuning, 39
tunneling-through processes, 171, *172*, *269*
tying up the years, 182

U.K. Wholesale Price Index, 176–77, *177*
uncertainties, expected, 37–38, 51, 234–35, *235–36*
United States:
    cars in, 154
    death rate in, 98, *99*, 101, *102*
    diphtheria in, 101–3, *102*
    energy use in, 114, 141, 147–55, *148*, *152–53*, 211–13, *212*, *256*, *266*
    future of, 92
    merchant vessels in, 133, *134*
    Nobel prizes and, 92, *93*, 94, 145
    nuclear energy in, 211–13, *212*
    oil imports of, 208–10, *210*
    overall personal "vehicle" market in, 110, *110*
    renewable energy in, 114, *256*
    steam engine in, 109, *255*
    substitution of detergent for soap in, 112, *113*, 114
Unsafe at Any Speed (Nader), 26
Utterback, James, 199

vaccines, 103
Valaoras, Vasilios, 30, 31
van der Erve, Marc, 126
van der Meer, Simon, 163
Van Gogh, Vincent, 156
VAX family products, 46–48
    MicroVAX II, 57, 59
    VAX 11/750, 57, *58*, 59, 234–35, *235–36*

VAX family products (*cont.*)
see also Digital Equipment Corporation
Verhulst, P. F., 55, 56
Verhulst dynamics, 196
Verhulst equation, 230
Viriakis, J., 220
vocabulary acquisition, 43, *43*, 130–31
Volterra, Vito, 11–12, 14–16, 55–56, 124
Volterra-Lotka system of differential equations, 55–56, 195, 196, 229–33
    fitting S-curve on set of data points and, 232–33
    Malthusian case and, 230–31
    multiple competition and, 232–33
    one-to-one substitutions and, 231–32
    predatory-prey system and, 229–30
volume curves:
    economies of scale and, 48
    S-curves and, 48–49

war:
    Heraclitus's view on, 124
    substitutes for, 183–85
    see also specific wars

water power, 139–40
weather forecasting, 20
Weiner, Ken, 141
Weingart, J., 214
Well-Tempered Clavier, The (Bach), 39
White, Newman Ivey, 82
Wholesale Price Index, U.K., 176–177, *177*
wind power, 139–40
women:
    birth rate of, 85–86, *248*
    as executives, 115–16, *257–58*
    Nobel prize and, *159*, 160
wood, 69, 139, *139*, 140, 142, 152
World War I, 104, 109
    death rate and, *99*
    diphtheria and, 100–104, *102*
World War II:
    death rate and, 98, *99*
    merchant vessels and, 133, *134*
    rubber and, 120, 210–11, *261*
    tuberculosis and, 104, *253*
writers, creativity/productivity of, 82–83, 86–87, *87, 246, 247, 249*

Zahavi, Yacov, 185